Strategic Sealift: Decisions Today To Ensure Tomorrow

NIMBLE BOOKS LLC: THE AI LAB FOR BOOK-LOVERS
~ FRED ZIMMERMAN, EDITOR ~
Humans and AI making books richer, more diverse, and more surprising.

Publishing Information

(c) 2024 Nimble Books LLC
ISBN: 978-1-60888-309-7

AI-generated Keyword Phrases

Commission on Merchant Marine and Defense; Strategic Sealift Plan; United States strategic sealift capabilities; Operations Desert Shield and Desert Storm; American President Companies, Ltd.; U.S. flag fleet charters; government and commercial companies; RRF assets; surge sealift requirements; additional strategic sealift assets; insufficient strategic sealift assets; 11 June 1991 publication date; contingency planning.

Publisher's Notes

Strategic sealift has never been more important to US national security, and US strategic sealift capability has never been weaker than it is today.

An ominous hint of what was to come can be found at the end of this 1991 thesis at the Naval War College, recommending:

> ... an intermodal network combined with Just In Time practices. U.S. flag carriers are seeing intermodalism as the watch-word of the future ... adapting an intermodal structure to strategic sealift requirements. Along the same lines, Just In Time provides today's businesses with up to the minute information and easy off-load access.

That worked out just great during the pandemic as global supply chains ground to a halt. Now, as the US prepares for war in the Pacific, it is finally sinking in that fighting wars overseas requires ample, independent, US-flagged strategic sealift.

In a world of escalating geopolitical tensions, the need to rapidly deploy forces and provide logistical support is more critical than ever. This book offers a compelling analysis of a nation's ability to project power across oceans and continents, connecting directly to the headlines of today. Beyond its strategic implications, the story of strategic sealift taps into our fundamental drive to explore, to venture beyond boundaries, and to ensure security and stability in a world often characterized by uncertainty.

This annotated edition illustrates the capabilities of the AI Lab for Book-Lovers to add context and ease-of-use to manuscripts. It includes several types of abstracts, building from simplest to more complex: TLDR (one word), ELI5, TLDR (vanilla), Scientific Style, and Action Items; essays to increase viewpoint diversity, such as Grounds for Dissent, Red Team Critique, and MAGA Perspective; and Notable Passages and Nutshell Summaries for each page.

ANNOTATIONS

Publishing Information ii
AI-generated Keyword Phrases ii
Publisher's Notes ii
Abstracts v
    Analysis Based on Full Context v
    Analysis Based on Abridged Content Windows vi
    TL;DR (one word) vi
    Explain It To Me Like I'm Five Years Old vi
    TL;DR (vanilla) vi
    Scientific Style vi
    Action Items vi
Viewpoints vii
    Grounds for Dissent vii
    Red Team Critique viii
    MAGA Perspective viii
Page-by-Page Summaries x
Notable Passages xv

ABSTRACTS

ANALYSIS BASED ON FULL CONTEXT

These analyses are created by using an LLM with a very long input context window, in this case Google Gemini 1.5-pro. The advantage is that the model can use the entirety of the document in its simulated reasoning.

This thesis investigates the capabilities of the United States Strategic Sealift program, which is responsible for transporting military personnel and equipment by sea during times of war or national emergency. The author argues that despite the apparent success of Operations Desert Shield and Desert Storm, the U.S. sealift capability remains insufficient and relies heavily on a confluence of fortunate circumstances during those operations.

The thesis provides a historical overview of the development of U.S. sealift capabilities, highlighting the crucial role of the Merchant Marine in supporting national defense. It examines various government studies conducted over the past few decades, all of which have consistently pointed to the inadequacy of existing sealift assets. The author argues that the success of Desert Shield and Desert Storm should not overshadow the underlying problems, emphasizing that factors such as the extended timeframe afforded by the Iraqi forces, favorable political climate, and robust support from allies contributed significantly to the mission's accomplishment.

The thesis explores proposed solutions to address the shortage of sealift assets, focusing on both government and commercial sector perspectives. It delves into topics such as the need for additional surge sealift vessels, revitalization of the Merchant Marine, and establishment of a Merchant Marine Reserve program. While acknowledging the merits of various proposals, the author emphasizes the need for a comprehensive approach involving cooperation and understanding between government agencies, the Navy, and commercial carriers to formulate effective solutions.

ANALYSIS BASED ON ABRIDGED CONTENT WINDOWS

These analyzes are created by analyzing portions or summaries of the document, using LLMs with shorter context windows. The advantage is that these models are faster and cheaper.

TL;DR (ONE WORD)

Sealift.

EXPLAIN IT TO ME LIKE I'M FIVE YEARS OLD

The government and commercial companies work together to make sure there are enough big ships to carry things like tanks and supplies in case we need them for an emergency or war. They have a plan to make sure there are enough ships ready to go when we need them. Sometimes they have to rent ships from other companies to help out. They are always working to make sure we have the ships we need to keep our country safe.

TL;DR (VANILLA)

The United States has taken steps to improve its strategic sealift capabilities, but still lacks enough assets to meet all potential contingencies. Both government and commercial companies have been involved in efforts to address this issue.

SCIENTIFIC STYLE

The Commission on Merchant Marine and Defense has identified a need for increased strategic sealift capabilities in the United States. The Strategic Sealift Plan outlines the shortage of assets for initial surges and contingency situations, prompting the chartering of ships from the U.S. flag fleet and the Ready Reserve Fleet (RRF). Despite efforts to address the deficiency, the U.S. still lacks sufficient strategic sealift assets. Recommendations have been made to bolster the country's sealift capabilities in response to potential future conflicts.

ACTION ITEMS

Increase funding for the Ready Reserve Fleet (RRF) to ensure sufficient strategic sealift assets are available in times of need.

Continue to charter ships from the U.S. flag fleet to supplement strategic sealift capabilities.

Collaborate with commercial companies to enhance strategic sealift capabilities and ensure readiness for any contingency.

Implement the recommendations of the Commission on Merchant Marine and Defense to strengthen the strategic sealift plan.

Conduct regular assessments and exercises to test and improve strategic sealift capabilities.

VIEWPOINTS

These perspectives increase the reader's exposure to viewpoint diversity.

GROUNDS FOR DISSENT

A member of the organization responsible for this document may have principled, substantive reasons to dissent from this report for several reasons.

Firstly, they may disagree with the assessment that the United States did not have enough assets to provide the required sealift, especially for the initial surge. They may argue that there were sufficient assets available or that alternative solutions could have been implemented to address any gaps in sealift capabilities.

Secondly, they may disagree with the assertion that much has been done to provide the United States with necessary strategic sealift capabilities. They may believe that more needs to be done to improve and expand sealift capabilities to ensure readiness for any contingency or war.

Additionally, the dissenting member may disagree with the statement that the United States remains with an insufficient number of strategic sealift assets despite the results of Operations Desert Shield and Desert Storm. They may argue that lessons learned from these operations should have been used to address any shortcomings in sealift capabilities.

Overall, the dissenting member may believe that the report's assessment of strategic sealift capabilities is inaccurate or incomplete, and that more needs to be done to address any gaps or deficiencies in order to ensure the United States is adequately prepared for any future contingencies.

RED TEAM CRITIQUE

The document appears to highlight the shortcomings in the United States' strategic sealift capabilities, particularly during times of crisis or war. The reliance on chartered ships from the U.S. flag fleet indicates a lack of sufficient assets within the government-owned Ready Reserve Fleet (RRF). This could potentially pose a significant risk in terms of the country's ability to rapidly deploy forces and resources in times of need.

From a government perspective, the critique would focus on the inadequacy of current strategic sealift assets and the need for increased investment in this area. The reliance on chartered ships may not always be a reliable or cost-effective solution, especially during times of heightened demand.

From a commercial company standpoint, the critique would likely highlight the potential opportunities for increased business in the maritime industry, particularly in the area of strategic sealift. Companies like American President Companies, Ltd. could potentially benefit from increased government contracts for sealift services.

Overall, the document emphasizes the importance of addressing the United States' strategic sealift capabilities to ensure readiness and preparedness for any future contingencies. The critique would likely highlight the need for a comprehensive and sustainable solution to address the gaps in strategic sealift assets.

MAGA PERSPECTIVE

This Commission on Merchant Marine and Defense Strategic Sealift Plan is just another example of government waste and overspending. The United States does not need to be pouring money into unnecessary assets for hypothetical contingencies. We should be focusing on America first, not wasting resources on unnecessary defense initiatives.

The idea that we need more strategic sealift capabilities is ridiculous. The United States is already the most powerful country in the world, we don't need to be constantly preparing for potential conflicts. This plan is just another example of the deep state trying to control every aspect of our military and defense strategy.

It is clear that the government and commercial companies involved in this plan are only looking out for their own interests. They are just trying to line their pockets with taxpayer money by pushing for unnecessary strategic sealift assets. This is a perfect example of the swamp that President Trump has been trying to drain since he took office.

The fact that this plan was put into place in 1991 and is still being discussed shows the inefficiency and lack of progress in our government. We need to be focusing on more important issues, like securing our borders and protecting American jobs, instead of wasting time and resources on outdated defense initiatives.

Overall, this Commission on Merchant Marine and Defense Strategic Sealift Plan is a perfect example of the deep state trying to control our military and defense strategy. We need to push back against these unnecessary and wasteful initiatives and focus on America first.

Page-by-Page Summaries

BODY-1   *Thesis on strategic sealift decisions for the future by Mary Alice Fults, advised by Dan C. Boger.*

BODY-2   *The page discusses the importance of strategic sealift for national security, highlighting the inadequacies in current capabilities and the need for decisive action to improve both government and commercial fleets.*

BODY-3   *A thesis on strategic sealift decisions for the future by Mary Alice Fults, submitted for a Master of Science in Management degree from the Naval Postgraduate School in June 1991.*

BODY-4   *Strategic Sealift is crucial for national security, but current capabilities are insufficient. Despite success in the Persian Gulf War, the US still needs an active Merchant Marine. Solutions are needed to improve government and commercial fleets for national security.*

BODY-5   *The table of contents outlines the motivation, scope, methodology, and organization of chapters on strategic sealift, Desert Shield/Desert Storm, and proposed solutions for insufficient ships, aging mariners, and declining shipyard industry.*

BODY-6   *The page provides conclusions and recommendations related to scenario planning, additional sealift, Ready Reserve Force readiness, prior investments, and recommendations for a Commission on Merchant Marine and Defense, Sealift Lobby in Washington, D.C., Maritime Administration, better understanding, sealift/transportation pipeline, strategic sealift plan, Army COSCOM onboard APF, and future study.*

BODY-7   *List of figures detailing the operation Desert Shield/Desert Storm, including U.S. and foreign flag charters, ships committed, RRF ships activated, performance, dry cargo and tanker delivery profiles, and dry cargo tons.*

BODY-8   *Dictionary of terms providing definitions for strategic sealift terminology, including APF (Afloat Prepositioning Fleet) and BB (Breakbulk ships), essential for understanding the subject material in the thesis.*

BODY-9   *Overview of EUSC ships, Flatracks, and LASH ships for military cargo transportation and readiness.*

BODY-10   *Military Prepositioned Squadron (MPS) and National Defense Reserve Fleet (NDRF) are fleets of ships maintained by the government for quick deployment in military operations. RO/RO ships are used for transporting combat equipment with fast turnaround capabilities. OPDS allows tankers to offload at sea.*

BODY-11   *Overview of Ready Reserve Force, SEA SHEDs, SEABEE Ships, Surge Sealift, and Sustainment Sealift in military logistics, detailing readiness status, equipment for cargo carrying, transportation of military supplies, and support for deployed forces in emergency situations.*

BODY-12   *The page discusses the potential scenarios for continued RRF activations, including U.S. and foreign flag charters and requisitioning if needed.*

BODY-13   *The author explores the importance of strategic sealift, highlighting the necessity of a strong Merchant Marine in times of conflict despite the success of existing fleets during Desert Shield and Desert Storm.*

BODY-14   *This page discusses the methodology and assumptions made in compiling information on U.S. Strategic Sealift, emphasizing the importance of a strong Merchant Marine for an effective program.*

BODY-15	*The page discusses the importance of improving surge sealift capabilities within the government's fleet, highlighting the challenges faced during Operations Desert Shield and Desert Storm due to deficiencies in the Merchant Marine.*
BODY-16	*Analysis of the initial surge in sealift assets during a crisis, highlighting the need for additional government ships and improvements in the Merchant Marine to ensure success.*
BODY-17	*Strategic Sealift has been a major mission area for the US Navy since 1985, but studies consistently show that the sealift capabilities are insufficient for national security needs.*
BODY-18	*Government studies from 1984 to early 1991 highlight ongoing sealift deficiencies, including insufficient strategic mobility capability and projected asset shortages, with findings validated by Operations Desert Shield and Desert Storm.*
BODY-19	*The page discusses the deteriorating condition of America's maritime industries and the historical context of insufficient sealift capabilities for accomplishing missions, emphasizing the need for strategic measures to address the situation.*
BODY-20	*The page discusses the challenges faced by the U.S. Merchant Marine during World War I and the measures taken to address the shortage of ships, including the enactment of the Shipping Act of 1916 to establish the United States Shipping Board.*
BODY-21	*The Merchant Marine Act of 1920 established U.S. shipping policy, aiming to sell off the government's fleet and develop a merchant marine owned and operated privately by U.S. citizens.*
BODY-22	*The Postal Act of 1928 required all U.S. mail to be transported on U.S. flag ships, leading to the Merchant Marine Act of 1936 which aimed to revive the American merchant marine industry through nationalistic policies.*
BODY-23	*US needed a strong merchant marine for defense and commerce, but struggled to meet demand in WWI and WWII. US flag fleet carried only 14% of world tonnage in 1939, increasing to 60% by end of WWII.*
BODY-24	*The page discusses the importance of a strong merchant marine for national security, highlighting the need for a large industrial base to support shipbuilding. It also mentions the post-WWII disposal of government ships and the impact on the industry.*
BODY-25	*The page discusses the role of the U.S. government in establishing the National Defense Reserve Fleet and the use of military sealift services during the Vietnam War to supply troops. It highlights challenges faced in logistics, including limited port infrastructure and civilian control issues.*
BODY-26	*Logistical challenges in Vietnam included lack of storage space for ships and fuel oil, leading to delays and disruptions. Despite these challenges, Strategic Sealift was able to handle all requirements during the war.*
BODY-27	*Decline of Merchant Marine industry impacted national security, highlighted by successful but ignominious withdrawal and lack of strategic sealift capability. Senator Denton's reports warned of deteriorating maritime industries and insufficient defense requirements. Desert Shield operation would have failed without key events.*
BODY-28	*Despite the success of Operations Desert Shield and Desert Storm, the United States lacks sufficient ships and personnel for full-scale contingency operations, relying on luck rather than preparedness. Warnings about declining Merchant Marine industry have been ignored.*

BODY-29	The Military Sealift Command is responsible for providing ocean transportation for all services and U.S. Government agencies during peacetime operations, with strategic sealift operations during war or national emergency situations. The order of call up for ships in a war or emergency situation is established.
BODY-30	Initial surge sealift for Desert Shield/Storm included deployment of Maritime Prepositioning Squadrons, Afloat Prepositioning Fleet, Fast Sealift Ships, and Ready Reserve Force ships, delivering equipment and supplies to the Persian Gulf by C + 8 and C + 14.
BODY-31	Rapid deployment of five U.S. divisions during Operation Desert Shield required activation of over 70 Ready Reserve Force ships, chartering of U.S. and foreign flag vessels, and continuous resupply efforts.
BODY-32	Foreign vessels chartered under SMESA agreement for U.S. cargo delivery to Persian Gulf faced issues with foreign control and laws, leading to refusal to enter war zones.
BODY-33	Cyprus maritime law allows seafarers to renegotiate or void contracts in case of war, unlike U.S. law where seafarers must stay on the ship. Foreign crews may not show up or rebel, causing issues during wartime operations. President Bush initiated Phase II of U.S. mobilization in November 1990.
BODY-36	The Ready Reserve Force ships were not activated within the expected timeframes during Desert Shield/Storm, leading to difficulties and challenges in the process.
BODY-38	Delays in activating Ready Reserve Fleet ships were caused by deteriorated gaskets, unfamiliar engineering personnel, crew shortages, and lack of maintenance during lay-up. Shipyard personnel shortages and unfamiliarity with foreign-built equipment also contributed to delays.
BODY-40	Budget cuts impacted maintenance of engineering plants on ships, requiring long distances traveled and modifications for hot, dusty Gulf environment. Ports had modern facilities due to Saudi modernization program. Heavy equipment arrived quickly after mobilization.
BODY-43	The success of the Coalition Armed Forces in the Gulf War was attributed to the critical factor of time, allowing for preparation and supply routes to be established before hostilities broke out.
BODY-45	The page discusses the impact of the Viet Nam conflict on the defensive and offensive strategies during the Iraq war, highlighting the importance of strategic sealift capabilities in ensuring successful operations.
BODY-46	Special Middle East Sealift Agreement (SMESA) was crucial for success, preventing strain on U.S. flag vessels and maintaining normal trade routes. Allied assistance with 85 chartered vessels was also vital, with a 0% attrition rate.
BODY-47	Saudi port facilities and Saudi-provided water and petroleum were crucial for efficient military operations during the Gulf War, saving on transportation costs and ensuring fast turnaround times for ships. The world shipping market for break bulk and roll on/roll off ships had slumped by the time of the conflict.
BODY-48	Decline in U.S. flag carriers' RO/RO ships due to economic reasons. U.S. military buildup under Reagan followed by drawdown in Europe in 1990. MSC anticipates Gulf War deployment with ships ready for transport.
BODY-49	Strategic sealift capabilities were tested during Operation Desert Shield, revealing the need for improvements in U.S. flag surge sealift and mariner availability. Success was partially due to fortuitous events, highlighting areas for further development in military sealift operations.

BODY-50   APL faced challenges in supporting Desert Shield/Desert Storm, risking loss of routes and customers. Urgent military lifts disrupted normal operations, leading to economic and logistical issues. US needs to address insufficient strategic sealift assets promptly.

BODY-51   Analysis of proposed solutions to insufficient Strategic Sealift assets, including need for additional surge assets and improvement in merchant marine industry. Consensus on importance of U.S. flag merchant marine for military efficiency. Concerns about potential conversion to foreign flag ships due to economic incentives.

BODY-52   Debate over subsidies and the definition of mobility base in relation to sealift assets in the U.S.-Flag Merchant Marine.

BODY-53   The Department of Defense is considering acquiring additional RO/RO vessels for surge sealift, but commercial container vessels have not yet been utilized for this purpose due to logistical challenges and the need for quick deployment of military assets.

BODY-54   DOD considering acquiring additional RO/RO vessels to increase surge sealift capacity, but building in U.S. shipyards and leasing impractical due to time and budget constraints. Better solution is to purchase ships from U.S. flag companies or foreign markets.

BODY-55   Recommendation to allow foreign-built United States flag ships to be eligible for ODS, with an initial year of worldwide procurement provision to address the need for more strategic sealift. Additional ships for the APF are also recommended to address Army's problems during conflicts.

BODY-56   Logistics issues in Vietnam War led to delayed support, not repeated in Desert Shield due to early deployment of Combat Support Command. Operating Differential Subsidy program criticized for subsidizing US flag vessels over foreign ones. Conflicting views on program's benefits.

BODY-57   Debate over Ocean Dumping Subsidies (ODS) in the US shipping industry - some argue for government support, others believe it hinders competitiveness and leads to reliance on subsidies for survival. Sea-Land advocates for eliminating ODS, stating it has led to a false sense of security.

BODY-58   Subsidies are crucial for the Merchant Marine industry, as they support national defense, economic welfare, and industrial base. Active ships are essential for reliable service, economic strength, and fair shipping costs. Reforms to existing legislation are needed until carriers can survive without subsidies.

BODY-59   Proposal for a Merchant Marine Reserve program to address shortage of trained crews, with questions about administration, training providers, and logistical challenges.

BODY-60   Proposing a merchant marine reserve program to increase readiness, but facing challenges in recruiting and training mariners due to strict Coast Guard requirements and modernization reducing crew needs.

BODY-61   Improving public relations for the Merchant Marine in civilian and Navy sectors can increase interest in maritime careers and foster better understanding and cooperation between the two groups.

BODY-62   Encouraging Navy sailors to pursue Merchant Marine licenses while on active duty can increase the number of qualified mariners and improve relations between the Navy and civilian mariners. The declining shipyard industry in the US is shifting focus from building to repairs.

BODY-63 Proposed bill aims to penalize ships built with subsidies after 1991, impacting shipowners' decisions. Shipbuilders Council of America lobbying to end shipbuilding subsidies worldwide. American shipbuilders urged to focus on pleasing commercial customers and finding profitable niches in international shipping market.

BODY-64 Revamping shipyard practices to improve quality and reputation, following Trinity Marine Group's lead. US shipyards need to focus on quality, marketing, and management to benefit their business. Increasing contracts for new ships at US shipyards would provide long-term work and maintenance opportunities.

BODY-65 Active ships are crucial for maintaining an industrial base, influencing international shipping rates, and contributing positively to the economy. The focus should be on asking the right questions to find effective solutions for strategic sealift needs.

BODY-66 Historical overview of strategic sealift issues, highlighting the inadequacy of existing assets and the need for more surge sealift capabilities, as demonstrated during Operation Desert Shield and Operation Desert Storm.

BODY-67 Insufficient U.S. flag sustainment sealift for various scenarios, with suggested solutions including old fears, new ideas, and unanswered questions. Conclusion: need for strategic sealift assets, changes in Merchant Marine, and clear definition of strategic sealift quantification. Lack of clear game plan for strategic sealift in Saudi Arabian area during Desert Shield/Storm.

BODY-68 Inadequate surge sealift for Desert Shield led to confusion and delays. Need for additional surge sealift is urgent, as demonstrated by history. Ready Reserve Force readiness is questionable. Prior investments in surge sealift in the 1980s were crucial for Desert Shield success.

BODY-69 Recommendations include reexamining findings of the Commission on Merchant Marine and Defense, making sealift a priority in Washington, D.C., strengthening the Maritime Administration's advocacy for the Merchant Marine, and promoting better understanding among key players in the industry.

BODY-70 Recommendations for improving cooperation and understanding within organizations, including the need for specialized training in sealift transportation and a strategic plan for sealift operations.

BODY-71 Recommendations for future study include exploring intermodal networks and Just In Time practices for Department of Defense strategic sealift, establishing agreements in volatile regions before hostilities, and examining the establishment of a Merchant Marine Reserve.

BODY-72 List of references related to military sealift programs and policies, including reports, theses, interviews, and government documents from various sources.

BODY-73 Various sources and interviews discussing the importance of the American Merchant Marine and logistical support in national security and military operations, including Operation Desert Shield/Desert Storm.

BODY-74 Discussions and correspondence with various maritime industry professionals and reports on the state of the industry in relation to military operations in the Persian Gulf.

BODY-75 Distribution list for copies of a document to various military and academic institutions.

NOTABLE PASSAGES

BODY-2 "Strategic Sealift is considered vital for our national security, and is often termed the 'Fourth Arm of Defense.' It is made up of two fleets, one owned and operated by the U.S. government, the other owned and operated by commercial companies and often chartered by the U.S. government. The most recent studies on the status of strategic sealift in the United States have all indicated that our present capabilities, in both fleets, are insufficient to handle anticipated national defense requirements."

BODY-4 The conclusion is that to ensure our national security the United States must take decisive action now to improve both the government and the Merchant Marine fleets.

BODY-8 "Contingency Operations => 'An emergency involving military force caused by natural disasters (such as volcanoes), terrorists; subversives or by required military operations. Due to the uncertainty of the situation, contingencies require plans, rapid response and special procedures."

BODY-9 "EUSC => 'Effective U.S. Controlled Fleet...(These) ships are considered requisitionable assets, available to the U.S. Government in time of national emergency. (They) are majority-owned by U.S. business, operated under the registries of four foreign nations - Liberia, Panama, Honduras and the Bahamas - and crewed by foreign nationals."

BODY-10 "The government pays the owners a set fee, leaving MSC free to do what it wants, when it wants, with any of the different ships. These ships are not useful for commercial purposes as the force consists of 13 RO/RO ships, split into three squadrons, and pre-loaded with enough equipment and supplies for each squadron to sustain an entire Marine Expeditionary Brigade (MEBs) for 30 days."

BODY-11 "Sustainment Sealift is the transportation of equipment, troops and supplies to continue the support of forward deployed forces in a war or other emergency situation."

BODY-13 "History, especially recent U.S. history, has proven the necessity for an active and able Merchant Marine in any conflict or emergency."

BODY-14 The need for a strong Merchant Marine, consisting of active and able mariners, ships that are militarily useful, and efficient, working shipyards, is paramount to an effective strategic sealift program.

BODY-15 "Improvements to the commercial fleet will help in sustainment sealift, improving our surge sealift capabilities can only be accomplished within the government's fleet."

BODY-16 Chapter IV will present and examine various solutions to the problem of insufficient sealift assets, delving into the areas of additional government ships and improvements in the Merchant Marine.

BODY-17 "Though for many years the need for a strong Strategic Sealift capability was consistently stressed at all levels, especially for its importance to national security, it was not added as a major mission until 1985, by then Secretary of the Navy John Lehman."

BODY-18 "The (Study) identified several sealift deficiencies that will exist throughout the 1990s. Included are shortages in unit equipment movement capacity through 1999; containerized cargo capacity shortages beginning in 1997; a 5-percent shortage of petroleum, oils, and lubricants capacity by 1999; decreasing utility of the commercial US-flag fleet to carry military cargo; a declining US shipbuilding industrial base; and a declining number of US merchant mariners." [Ref. 7:Encl (3)] Most of these problems already exist;

BODY-19 "During their two year investigation, the Commission never encountered anything to dispel this finding. These conditions however, did not come into existence overnight. It has taken many years and many events for the number and expertise of mariners to decline to its present state, the shipbuilding industry to fall to such disrepair, and the number of active ships to shrink to a modem day low."

BODY-20 "The U.S. Merchant Marine was also at an all time low, with trained mariners difficult to find. The continued reliance on foreign ships to transport our export and import trade goods resulted in an economic crisis of a few ships charging astronomical rates."

BODY-21 "It is necessary for the national defense and for the proper growth of its foreign and domestic commerce that the United States shall have a merchant marine of the best equipped and most suitable types of vessels sufficient to carry the greater portion of its commerce and serve as a naval or military auxiliary in time of war or national emergency, ultimately to be owned and operated privately by citizens of the United States; and it is hereby declared to be the policy of the United States to do whatever necessary to develop and encourage the maintenance of such a merchant marine."

BODY-22 "An American merchant marine is one of our most firmly established traditions. It was, during the first half of our national existence, a great and growing asset. Since then, it has declined in importance and value. The time has come to square this traditional ideal with effective performance." - President Roosevelt, Merchant Marine Act of 1936

BODY-23 "It is necessary for the national defense and development of its foreign and domestic commerce that the United States shall have a merchant marine (a) sufficient to carry its domestic waterborne commerce and a substantial portion of the waterborne export and import foreign commerce of the United States and to provide shipping service on all routes essential for maintaining the flow of such domestic and foreign waterborne commerce at all times."

BODY-24 "The very fact that we needed such an undertaking is a warning that, at least for national security purposes, the United States must have a strong merchant marine at all times. There is another warning inherent within this piece of history - only by having a large industrial base within our shipyard industry, were we capable of producing over 5,000 ships in a three-year time span."

BODY-25 "In total, 95 percent of all military cargo delivered to South Vietnam was transported by strategic sealift."

BODY-26 "The average time a deep draft ship waited for a berth in Vietnam ports (went) from 20.4 days during the most critical period of 1965 to the 1970 average of less than two days."

BODY-27 "In its first report the Commission concluded that there is a 'clear and growing danger to the nation's security in the deteriorating condition of America's maritime industries.' Senator Denton continued in his fourth, and final report: 'Both our strategic sealift capability and our shipyard mobilization base today fall significantly short of defense requirements, and we (the commissioners) fear that without decisive action the situation will worsen substantially by the year 2000.'"

BODY-28 "Today, even with the Fast Sealift Ships, the Ready Reserve Force (RRF) and the many chartered ships in the government's fleet, the United States does not have a sufficient level of the necessary (militarily useful) ships and qualified personnel to handle a full scale contingency operation."

BODY-29   "When the President's call went out, MSC was the one to answer for strategic sealift operations."

BODY-30   "The initial surge sealift was provided by launching two of the three Maritime Prepositioning Squadrons (MSPRONs), ten ships from the Afloat Prepositioning Fleet, seven of the Fast Sealift Ships and an activation call for ships from the Ready Reserve Force. Due to the MSPRONs deploying from Diego Garcia (British Indian Ocean Territory), Guam (Western Pacific) and Saipan (Western Pacific/North China Sea) the first heavy combat equipment was delivered to the Gulf by C + 8 (15 Aug); a total of 105,000 short tons."

BODY-31   Ultimately, five entire U.S. Divisions were be sent (though all previous plans had been for a total of three and one-third divisions), with all the assorted necessary equipment and supplies.

BODY-32   "In one instance, the officers and crew of a feeder ship chartered in by American President Lines refused to take the vessel into the Persian Gulf area once war was declared. Interestingly, though documents filed with MarAd indicated that the vessel was registered in Cyprus, its registration had expired. The documents filed with MarAd indicated that the ship had German officers and a Maltese crew. Had the vessel been registered in Cyprus, these civilian mariners would have been legally entitled to refuse to enter a war zone."

BODY-33   "In another incident, the 27 Bangladeshi crew members of the freighter Banglar Mamatar quit lining the holds of the vessel with plywood in Oakland, Calif., prior to loading military cargo for the Gulf, and jumped ship.... Reportedly, Sea-Land also had some problems just prior to hostilities when the crew of one foreign-flag ship refused to enter the Gulf."

BODY-36   "In deference to the Maritime Administration, this five, ten or 20-day requirement is simply an arbitrary number assigned to each ship. There are no clear guidelines for determining just how long it should take to reactivate a ship."

BODY-38   "The most consistent problem encountered was also the worst; the condition of the vessel at breakout. If the ship had been in lay-up for an extended period of time, with no maintenance activation conducted, it was inevitable that major problems would be experienced throughout the vessel."

BODY-40   The environment of the Gulf is hot and extremely dusty. This requires many modifications to normal shipboard operations, including special protection for ventilation ducts, and careful monitoring of temperatures in the engineering plants. Additionally, heat stroke and sunburn are constant dangers.

BODY-43   By not attacking the coalition forces, Saddam Hussein gave the U.S. and her allies time to prepare. Time was the critical factor.

BODY-45   "The planning and expenditures that went into strategic sealift capability over the last decade have paid off handsomely. The kind of ships we acquired for contingency sealift operations have done the job we hoped they would do." - Vice Adm. Francis R. Donovan, USN

BODY-46   "The cooperation achieved was, to say the least, phenomenal; without that agreement serious shortfalls might have been felt. While the U.S. flag vessels were available, if they had had to make all the deliveries, their own regular trade routes would have suffered. This would have put a serious strain on, not only the carrier's economic status, but also on the nation's."

BODY-47   "The amazing and modern Saudi port facilities, built in the early 80's, were also a phenomenal blessing. They provided the ability for all the ships to make a fast turnaround, as well as have a place to store the delivered material along with the containers in which much of the material was delivered."

BODY-48 During President Reagan's terms in office, the U.S. Armed Forces experienced the greatest buildup since WWII. This occurred in spite of the 1984 Gramm/Rudman/Hollings Budget Act. Then, in late 1989 and throughout 1990, changing strategies within the United Soviet Socialist Republics caused the USSR to experience what can be termed 'major setbacks' in Eastern Europe. This has led to the United States setting back its own time tables (from a few days expected warning to many months) and essentially relaxing its defensive posture.

BODY-49 "We should not leave this exercise and say our strategic lift is OK, because it isn't. We're a victim of our success." - Vice Admiral Paul D. Butcher.

BODY-50 "The bottom line is that the United States is still sitting with insufficient strategic sealift assets. Whatever action is to be taken to change this situation needs to be decided now, especially in light of the recent Persian Gulf events."

BODY-51 Based on its study and analysis, the Commission has concluded that there is no more militarily efficient, cost effective, and reliable way to provide the majority of the military sealift requirement now and in the future than through an active United States flag merchant marine. The ships should be militarily useful and operating, engaged in peacetime in carrying commercial cargo, and manned by United States crews.

BODY-52 Sea-Land, claims that "(w)e are a strong advocate to dispose of ODS (Operating Deficiency Subsidy) because we think it has contributed to the demise of the U.S.-Flag Merchant Marine."

BODY-53 (W)e do believe that jointly, the DOD and Industry will find a way for the container vessels and the intermodal infrastructures to contribute to the surge phase of strategic sealift to a much larger degree than experienced in Desert Shield/Storm.

BODY-54 "The better solution would be to purchase these ships, either from the U.S. flag companies ready to sell or in foreign markets. This would make the ships available now."

BODY-55 The recommendation allowing foreign-built United States flag ships to be eligible for ODS should ultimately have no limits in terms of time, ship type, and numbers, but the Commission recognizes that, realistically, such a provision is not now desirable. The Commission therefore recommends that, for an initial year, the worldwide procurement provision be implemented, in parallel with the institution of the domestic shipbuilding programs recommended elsewhere in this report to expand the size of the militarily useful fleet.

BODY-56 "Stemming from years of controversy, poor management and corrupt administrators, any subsidy is considered bad. The ODS program is no exception. It is basically a means for subsidizing the difference in cost between operating a U.S. flag vessel with U.S. unionized crews and a foreign vessel with foreign crews."

BODY-57 "We are a strong advocate to dispose of ODS because we think it has contributed to the demise of the U.S.-Flag Merchant Marine."

BODY-58 - Active ships are immediately operational and capable of providing reliable service;

- Active ships help maintain the active industrial base needed to support an expanded strategic sealift force during time of war or national crisis;

- Active ships contribute to American economic strength;

- Active ships provide the United States with a larger voice in international rate-making conferences and ensure that foreign shipping cartels will not unfairly manipulate shipping costs;
- Active ships in a healthy United States merchant marine pay taxes and contribute favorably to the balance of payments.

BODY-59 "A Merchant Marine Reserve program is one of the methods being examined as a possible solution to the diminishing numbers of licensed and unlicensed marine personnel. The idea behind this program is to provide a core of mariners who would then train periodically. Then, just as the service reserves are on call during an emergency situation, so too would the merchant marine reserves be available."

BODY-60 "In other words, provide a reason to join the merchant marine reserve vice one of the service reserves.. Additionally, a merchant marine reserve might end up working contrary to the basic idea of increasing the number of seafarers."

BODY-61 "With a concentrated effort to improve relations between the Navy and the civilian mariners, each side could see better service and understanding from, and towards, the other."

BODY-62 Active ships provide a cadre of trained merchant seamen to crew both active and/or reserve ships; Active crews are trained and familiar with the equipment and operational requirements of the ships; Active ships in a healthy United States merchant marine pay taxes and contribute favorably to the balance of payments.

BODY-63 "American shipbuilders have to relearn what's involved in pleasing commercial customers." [Ref. 35:p 101]

BODY-64 "Trinity's 'secret' is to provide the best price, quality, and delivery," say spokesman A.J. Rizzo. "The company has an aggressive sales team that hears about work and doesn't stop until they get it. Trinity is also flexible enough to take on anything, including large deep-sea vessels at its Beaumont, Texas, facility."

BODY-65 "To paraphrase Admiral Robert J. Kelly, Commander-in-Chief U.S. Pacific Fleet, 'more important than finding the right answers is to find the right questions.' Prior to solving the problem of insufficient strategic sealift, the right questions must be found and asked. Only then can the right answers also be found and implemented."

BODY-66 "The most important lesson learned from Operation Desert Shield, though, was that the United States does not have enough surge sealift to meet all Department of Defense requirements."

BODY-67 The first and obvious conclusion has already been made. Despite the results of Operations Desert Shield and Desert Storm, the United States remains with an insufficient number of strategic sealift assets. The second basic conclusion is that if changes to the Merchant Marine are to take place they need to happen now, before the industry reaches a point of no return. The third is that there exists no clear definition on how strategic sealift is to be quantified, whether by ship size, by number of ships or by cargo capacity. Before any other decisions on acquisitions can be made, the government needs to know how much to acquire.

BODY-68 "If a national emergency can spring up with the suddenness of Desert Shield, five days from invasion to call-up, then waiting to acquire more sealift is ignoring history."

BODY-69 "It is recommend that a close reexamination of the Commission's findings of fact, conclusions and recommendations be made immediately. These four reports are invaluable in their content and deserve a concerted effort toward implementing their recommendations."

BODY-70 "It is recommend that the U.S. Navy examine the possibility of a program specifically designed to educate a cadre of personnel in the area of sealift transportation, for eventual use in a sealift or transportation designator. While the Transportation Management curriculum at the Naval Postgraduate School in Monterey, California, is excellent, it is specifically designed to fill certain P-coded billets, rather than be a pipeline to a transportation designator."

BODY-71 The use, by the Department of Defense, of an intermodal network combined with Just In Time practices. U.S. flag carriers are seeing intermodalism as the watchword of the future. A study should be made into the possibility of adapting an intermodal structure to strategic sealift requirements. Along the same lines, Just In Time provides today's businesses with up to the minute information and easy off-load access. This too should be studied for adaptability in parallel to the former recommended study.

AD-A246 071

NAVAL POSTGRADUATE SCHOOL
Monterey, California

DTIC
ELECTE
FEB 20 1992
D

THESIS

Reproduced From
Best Available Copy

STRATEGIC SEALIFT:
DECISIONS TODAY TO ENSURE TOMORROW

by

Mary Alice Fults

June 1991

Thesis Advisor: Dan C. Boger

Approved for public release; distribution is unlimited

92-04001

Unclassified
Security Classification of this page

REPORT DOCUMENTATION PAGE

1a Report Security Classification Unclassified	1b Restrictive Markings
2a Security Classification Authority	3 Distribution Availability of Report
2b Declassification/Downgrading Schedule	Approved for public release; distribution is unlimited
4 Performing Organization Report Number(s)	5 Monitoring Organization Report Number(s)

6a Name of Performing Organization Naval Postgraduate School	6b Office Symbol	7a Name of Monitoring Organization Naval Postgraduate School
6c Address (city, state, and ZIP code) Monterey, CA 93943-5000		7b Address (city, state, and ZIP code) Monterey, CA 93943-5000
8a Name of Funding/Sponsoring Organization	8b Office Symbol	9 Procurement Instrument Identification Number

8c Address (city, state, and ZIP code)	10 Source of Funding Numbers			
	Program Element Number	Project No	Task No	Work Unit Accession

11 Title (Include Security Classification)
STRATEGIC SEALIFT: DECISIONS TODAY TO ENSURE TOMORROW

12 Personal Author(s)
Fults, Mary Alice

13a Type of Report Master's Thesis	13b Time Covered From To	14 Date of Report (year, month, day) June 1991	15 Page Count 76

16 Supplementary Notation The views expressed in this thesis are those of the author and do not reflect the official policy or position of the Department of Defense or the U.S. Government.

17 Cosati Codes			18 Subject Terms (continue on reverse if necessary and identify by block number)
Field	Group	Subgroup	Strategic Sealift Military Sealift Command

19 Abstract (continue on reverse if necessary and identify by block number)
Strategic Sealift is considered vital for our national security, and is often termed the "Fourth Arm of Defense." It is made up of two fleets, one owned and operated by the U.S. government, the other owned and operated by commercial companies and often chartered by the U.S. government. The most recent studies on the status of strategic sealift in the United States, have all indicated that our present capabilities, in both fleets, are insufficient to handle anticipated National defense requirements. This thesis is an investigation into our capabilities in light of the recent Persian Gulf war. Some decision makers in Washington are saying that, due to the outstanding results of Operations Desert Shield and Desert Storm, the United States no longer needs an active Merchant Marine. Despite these results our "Fourth Arm" is still insufficient. This thesis examines the reasons why this is true, and concidters possible solutions to this problem, some of which have been provided from both the government and commercial companies. The conclusion is that to ensure our national security the United States must take decisive action now to improve both the government and the Merchant Marine fleets.

20 Distribution/Availability of Abstract [X] unclassified/unlimited [ ] same as report [ ] DTIC users.	21 Abstract Security Classification Unclassified	
22a Name of Responsible Individual Dan Boger	22b Telephone (Include Area code) (408) 646-2607	22c Office Symbol AS/BO

DD FORM 1473, 84 MAR 83 APR edition may be used until exhausted security classification of this
All other editions are obsolete Unclassified

i

Approved for public release; distribution is unlimited.

Strategic Sealift:

Decisions Today to Ensure Tomorrow

by

Mary Alice Fults
Lieutenant, United States Navy
B. F. A., University of Arizona

Submitted in partial fulfillment of the requirements
for the degree of

MASTER OF SCIENCE IN MANAGEMENT

from the

NAVAL POSTGRADUATE SCHOOL
June 1991

Author: _____
Mary Alice Fults

Approved by: _____
Dan C. Boger, Thesis Advisor

_____
Alan W. McMasters, Second Reader

_____ for
David R. Whipple, Chairman
Department of Administrative Sciences

ABSTRACT

Strategic Sealift is considered vital for our national security, and is often termed the "Fourth Arm of Defense." It is made up of two fleets, one owned and operated by the U.S. government, the other owned and operated by commercial companies and often chartered by the U.S. government. The most recent studies on the status of strategic sealift in the United States have all indicated that our present capabilities, in both fleets, are insufficient to handle anticipated National defense requirements. This thesis is an investigation into strategic sealift capabilities in light of the recent Persian Gulf war. Some decision makers in Washington are saying that, due to the outstanding results of Operations Desert Shield and Desert Storm, the United States no longer needs an active Merchant Marine. Despite these results our "Fourth Arm" is still insufficient. This thesis examines the reasons why this is true and considers possible solutions to this problem, some of which have been provided by the government and commercial companies. The conclusion is that to ensure our national security the United States must take decisive action now to improve both the government and the Merchant Marine fleets.

TABLE OF CONTENTS

I. INTRODUCTION 1
   A. MOTIVATION OF THE PROBLEM..1
   B. SCOPE..2
   C. METHODOLOGY 2
      1. Assumptions 2
   D. CHAPTER ORGANIZATION...3

II. BACKGROUND 5
   A. STRATEGIC SEALIFT 5
      1. Recent Studies 5
   B. EMERGENCE AND IMPORTANCE OF STRATEGIC SEALIFT..........7
      1. World War I 7
      2. World War II...11
      3. Vietnam War 13
   C. CONCLUSION 15

III. DESERT SHIELD/DESERT STORM..16
   A. SUCCESSFUL, BUT--...16
      1. Analysis 16
      2. The Facts of Desert Shield...16
      3. What Made It Successfull?..31
      4. Conclusions 37

IV. PROPOSED SOLUTIONS 39
   A. DIFFERENCES IN THOUGHT 39
      1. Solving the Problem of Insufficient Ships....................................40
      2. Insufficient and Aging Mariners...46
      3. Declining Shipyard Industry 50
   B. IN CONCLUSION 53

CONCLUSIONS AND RECOMMENDATIONS 54
    A. SUMMARY 54
    B. CONCLUSIONS 55
        1. Scenario Planning 55
        2. Additional Sealift 56
        3. Ready Reserve Force Readiness 56
        4. Prior Investments 56
    C. RECOMMENDATIONS 57
        1. Commission on Merchant Marine and Defense 57
        2. Sealift Lobby in Washington, D.C. 57
        3. Maritime Administration 57
        4. Better Understanding 57
        5. Sealift/Transportation Pipeline 58
        6. Strategic Sealift Plan 58
        7. Army COSCOM onboard APF 58
        8. Future Study 59

LIST OF REFERENCES 60

INITIAL DISTRIBUTION LIST 63

LIST OF FIGURES

3.1	OPERATION DESERT SHIELD/DESERT STORM U.S. AND FOREIGN FLAG CHARTERS	22
3.2	SHIPS COMMITTED TO DESERT SHIELD	23
3.3	OPERATION DESERT SHIELD/DESERT STORM RRF SHIPS ACTIVATED/TENDERED TO MSC	25
3.4	Operation Desert Shield/Desert Storm RRF Performance	27
3.5	DESERT STORM SEALIFT DRY CARGO DELIVERY PROFILE	29
3.6	DESERT STORM SEALIFT TANKER DELIVERY PROFILE	30
3.7	DRY CARGO TONS OPERATION DESERT SHIELD/DESERT STORM	32

DICTIONARY OF TERMS

In writing this thesis, the author has made the assumption that the reader will have a strong background in the subject material. However, since not all people start with the same foundation or knowledge base, a dictionary of terms is herein provided. This dictionary includes both strategic sealift terminology and their definitions.

APF => Afloat Prepositioning Fleet. Positioned in the Indian Ocean and Mediterranean since 1980 and similar in scope to the MPS, the only difference between the two groups is that the APF ships are pre-loaded with Army, Air Force and Navy supplies and equipment rather than those of the Marine Corps. The Military Sealift Command (MSC) currently charters 11 of these ships. [Ref. 1:p 19]

BB => "Breakbulk ships. This is the largest category of ships within the (RRF). . . . (These ships) are labor intensive and have long load and off-load times. The advantage of breakbulk ships is their self-sustainability, the ability to discharge cargo offshore by use of ships' booms and cranes. They are also capable of handling most military cargoes. The breakbulk's are generally faster ships with speeds in excess of 20 knots and (are steam driven). (Their capacity) is about 12,000 to 14,000 (dwt)." [Ref. 2:pp 36-37]

Contingency Operations => "An emergency involving military force caused by natural disasters (such as volcanoes), terrorists, subversives or by required military operations. Due to the uncertainty of the situation, contingencies require plans, rapid response and special procedures to ensure the safety and readiness of personnel, installations and equipment." [Ref. 3:p 1]

EUSC => "Effective U.S. Controlled Fleet. . .(These) ships are considered requisitionable assets, available to the U.S. Government in time of national emergency. (They) are majority-owned by U.S. business, operated under the registries of four foreign nations -- Liberia, Panama, Honduras and the Bahamas -- and crewed by foreign nationals. These countries, unlike most others, do not have laws which preclude or limit requisitioning. The EUSC ships number over 400, but only 23 dry cargo ships and 57 tankers are considered useful for military purposes. Manning with U.S. citizen crews may be required in certain circumstances." (All as of April 1985) [Ref.1:p 9]

Flatracks => Part of the Sealift Enhancement Features (SEF), they are selective equipment for commercial vessels to increase military cargo carrying capability. "(They) provide a capability to use containerships to carry oversize cargo. They expand the usefulness of commercial containerships in the rapid movement of military cargo, particularly wheeled/tracked vehicles. Flatracks are portable, open-topped, open sided units which fit into existing below-deck container cell guides. [Ref. 1:pp 14 and 39]

LASH => Lighter Aboard SHip. LASH ships are used in sustaining military supplies or carrying unit equipment. They operate in a manner similar to the container ship, lifting the lighters or barges out of the water by means of an overhead, traveling gantry crane which will then stack the lighter atop other lighters in a cargo cell. [Ref. 2:p 36]

MPS or MPSRON => Military Prepositioned Squadron: Ships under five to ten year charters with options for long-term charters up to 20 years. The government pays the owners a set fee, leaving MSC free to do what it wants, when it wants, with any of the different ships. These ships are not useful for commercial purposes as the force consists of 13 RO/RO ships, split into three squadrons, and pre-loaded with enough equipment and supplies for each squadron to sustain an entire Marine Expeditionary Brigade (MEBs) for 30 days. Their primary mission is to be available at a moment's notice for any operation requiring one or more MEBs. [Ref. 4, 5:p 13]

NDRF => National Defense Reserve Fleet: The NDRF is a fleet of ships, under the control of the Maritime Administration, "maintained in a condition that permits activation in one to six months." [Ref. 6:p 3]

OPDS => Offshore POL (Petroleum, Oil, and Lubricants) Discharge System. This system allows a tanker to offload while sitting in the stream rather than pier-side. It also allows other tankers to off-load through a buoy and hose line to the shore. [Ref. 7]

RO/RO => Roll on/Roll off. ". . . these ships are used for the . . . movement of oversized combat equipment. They have the distinct advantage of fast turnaround as moving vehicles can be driven (up and) down their ramps. They normally require a developed port to discharge their cargo; however, the Navy has developed a system for use in low seas that enables vehicles to be driven onto lighterage. Most . . . are diesel powered and are capable of carrying about 20,000 to 30,000 deadweight tons (dwt) of cargo at a speed of about 21-23 knots." [Ref. 2:p 36]

RRF => Ready Reserve Force: In contrast to the NDRF, the RRF ships are maintained in five-, ten- and 20-day readiness status. [Ref. 6: p 3]

SEA SHED=> Part of the Sealift Enhancement Features (SEF), they are selective equipment for commercial vessels to increase military cargo carrying capability. The "SHEDs provide temporary decks in containerships for transport of large military vehicles and outsize breakbulk cargo that will not fit into containers. Each SEA SHED is a structure (40'L x 24'W x 12.5'H) which fits into the cells of a containership occupying the space of four and one-half containers. The ship's load-bearing container cell guides must be reinforced before SEA SHEDs can be installed." [Ref 1:pp 14 and 38]

SEABEE Ships => "These ships are used in sustaining military supplies or carrying unit equipment. (T)he lighters are lifted by means of an elevator and are moved to different deck levels where they are transported forward for securing. There is no height limitation placed on the cargo in a lighter. SEABEE ships carry 38 1,000-ton capacity barges which are loaded by a stern elevator." [Ref. 2:p 36]

Surge Sealift => The initial transportation of equipment, troops and supplies to support deployed forces in a war or other emergency situation. Standard senario would be a Carrier Battle Group, followed by the Air Landed/Force entry, then the MPS, the APF, the Fast Sealift and finally the RRF/U.S. flag charters. [Ref. 8]

Sustainment Sealift => Follow-on to surge sealift, sustainment is the transportation of equipment, troops and supplies to continue the support of forward deployed forces in a war or other emergency situation. Typical senario would be for continued RRF activations, U.S flag and foreign flag charters and, if necessary, requisitioning.

I. INTRODUCTION

A. MOTIVATION OF THE PROBLEM

With the invasion of Kuwait by Iraqi troops it became obvious that this author thesis would be in the area of strategic sealift. In-depth research was begun almo immediately into this chosen area. By the time the war had reached its qui conclusion, numerous facts, reports and studies had been read and analyzed by th author.

Throughout the aforementioned research, it became clear that sufficie strategic sealift was an age-old problem. For the United States this has be especially true during the last century. Equally evident was the vital role played all aspects of the Merchant Marine. It was with great concern, therefore, that th author heard of a growing movement within Washington. *Since Desert Shield a Desert Storm were so successfully supported by the existing U.S. fleets, t government should not put forth any effort, or funds, to assist the Merchant Mari out of its current decline. Instead all funds should go towards increasing t government's fleet of prepositioned and fast sealift ships.*

This author takes exception to the above on two points. One; history, especial recent U.S. history, has proven the necessity for an active and able Merchant Marii in any conflict or emergency. Two; the successful sealift of Operations Desert Shie and Desert Storm was more the result of fortuitous events than anything else.

B. SCOPE

Strategic Sealift is a vast subject with many nuances. Numerous reports and studies have already covered its many aspects, far more in-depth than this thesis could hope to accomplish. Therefore, instead of "re-inventing the wheel," this thesis compiles, collates and briefly states, not only the significant results of the previous reports, but also the background for U.S. Strategic Sealift. The emphasis will then be placed on the solutions to our strategic sealift problem, as seen by the government and by the commercial sector.

C. METHODOLOGY

The methodology for this thesis was a study of various published and unpublished papers. These included periodicals, textbooks, Government studies, Naval Postgraduate School theses, and briefing notes and slides. Additionally, the author conducted phone and personal interviews, some of which culminated in facsimile traffic from the government and commercial sources. The applicable information was then examined, analyzed and collated into a comprehensive report on the subject. Whenever possible, the most current information available was obtained and used in this paper.

1. Assumptions

In writing this thesis some important assumptions were made:

a. The need for a strong Merchant Marine, consisting of active and able mariners, ships that are militarily useful, and efficient, working shipyards, is paramount to an effective strategic sealift program.

b. The best method for ensuring a sufficient amount of resources for strategic sealift is not by doing away with the active commercial fleet, but by a balance of an improved commercial fleet and additional pre-positioned and reserve ships.

c. While improvements to the commercial fleet will help in sustainment sealift, improving our surge sealift capabilities can only be accomplished within the government's fleet.

D. CHAPTER ORGANIZATION

Chapter II will provide a brief history of strategic sealift within the United States, proving the first point listed in section A above. This will be accomplished by a review of previous government studies, along with key points in American history. Because of the many reports and studies already covering this topic, only short synopses of them will be provided.

Chapter II will present an analysis of the strategic sealift aspects of Operations Desert Shield and Desert Storm. This will be done in order to prove the second point listed in section A above, along with its corollary: If the strategic sealift success of Desert Shield/Desert Storm was due in large part to fortuitous events, then the reported problems of the deficiencies and continued decline of the Merchant Marine still exist.

The analysis will cover the last months of 1990 and the early months of 1991, during which the United States faced a major crisis of global proportions. Under the auspices of the Military Sealift Command (MSC), the Navy was tasked with ensuring that enough assets were available to meet initial surge and then sustainment operations. However, the Navy has never been expected to meet these requirements with just the government's fleet. Instead, the main source of our strategic sealift is supposed to be, *as it has been legislated to be*, the active, U.S. flag, commercial fleet. In the Persian Gulf crisis, though, MSC came up against a major problem stemming from years of decline within the Merchant Marine. The United States did not have enough assets to provide the required sealift, especially for the initial surge. The conclusion is that only due to fortuitous events did the over[all] sealift succeed.

Chapter IV will present and examine various solutions to the problem [of] insufficient sealift assets, delving into the areas of additional government ships a[nd] improvements in the Merchant Marine. These solutions will be taken fro[m] government sources, major U.S. flag companies, as well as this author's own ideas.

Chapter V will bring everything together with a summary of the thes[is] conclusions and recommendations.

II. BACKGROUND

A. STRATEGIC SEALIFT

On August 1, 1990, the United States Navy had four major mission areas: Power Projection, Sea Control, Strategic Deterrence and Strategic Sealift. Though for many years the need for a strong Strategic Sealift capability was consistently stressed at all levels, especially for its importance to national security, it was not added as a major mission until 1985, by then Secretary of the Navy John Lehman. Before <u>and after</u> its inclusion, however, every study and report regarding our sealift capabilities has claimed that they were, and remain, insufficient.

1. Recent Studies

Each of these studies was classified at a Confidential level, or higher. What follows here are short, unclassified synopses of their findings. These brief descriptions were provided to the author by Military Sealift Command (MSC) after their compilation for the most recent Congressionally-mandated study of 1991.

a. *Government Study, 1981*

"The objective of the study was to determine an acceptable US mobility response capability within affordability limits." [Ref. 7:Encl (3)] Projecting out to 1986 levels, the report stressed that the existing lift capabilities were insufficient to support the 'current war-fighting strategies.' Many of the recommendations made by this study were used during the military buildup times of the 1980's. [Ref. 7:Encl (3)]

b. *Government Study, 1984*

Extensive in its coverage, though scenario-specific in its model, "(t)his study stated that the capacity may exist to meet sealift objectives in terms of amount and types of cargo, but it cannot meet the required delivery date for particular units." [Ref. 7: Encl (3)] When overlaid with the recent results of Operations Desert Shield and Desert Storm, the findings of this study are borne out as being 'right on the mark.'

c. *Government Study, 1989*

Looking at global deployment, and even without any 'fiscal constraints,' this study concluded "that the FY 1992 strategic mobility capability was insufficient to meet required delivery dates." Again Operations Desert Shield and Desert Storm provided clear evidence on the truth of these findings. [Ref. 7:Encl (3)]

d. *Government Study, 1990*

This study identified a "significant sealift shortfall," claiming that, by the year 2005, "projected US-controlled sealift assets would be insufficient to close desired forces on time." [Ref. 7:Encl (3)]

e. *Government Study, early 1991*

"The (Study) identified several sealift deficiencies that will exist throughout the 1990s. Included are shortages in unit equipment movement capacity through 1999; containerized cargo capacity shortages beginning in 1997; a 5-percent shortage of petroleum, oils, and lubricants capacity by 1999; decreasing utility of the commercial US-flag fleet to carry military cargo; a declining US shipbuilding industrial base; and a declining number of US merchant mariners." [Ref. 7:Encl (3)] Most of these problems already exist;

the expectation is that without drastic measures the situation will only get worse.

f. *Commission on Merchant Marine and Defense, 1987-88*

Though not included in the compilation of studies provided by MSC, the four reports submitted to the President by this Commission are just as relevant. (As the investigations and studies performed were not scenario dependent, each of these reports is unclassified and open to the public for review.) In their first report, *Findings of Fact and Conclusions*, the Commission found "clear and growing danger to the national security in the deteriorating condition of America's maritime industries." During their two year investigation, the Commission never encountered anything to dispel this finding. [Ref. 9:cover letter] These conditions however, did not come into existence overnight. It has taken many years and many events for the number and expertise of mariners to decline to its present state, the shipbuilding industry to fall to such disrepair, and the number of active ships to shrink to a modern day low.

B. EMERGENCE AND IMPORTANCE OF STRATEGIC SEALIFT

While the recent studies have all cited our insufficient sealift capabilities for accomplishing specified and general missions, this problem is, in actuality, nothing new to the United States.

1. World War I

Prior to WWI, President Woodrow Wilson stood in front of the Congress and declared as long as the nation depended on foreign ships for transport, "our merchants are at their (foreign nations) mercy, to do with as they please." He was speaking for a strong, economically viable and, most importantly, a national fleet independent of foreign shipping. [Ref. 10:p 23] At that time only about nine percent of American cargoes were being carried in U.S. flag ships. [Ref 11:p 23]

At the outbreak of hostilities in Europe, the United States found itself with too few ships to cover the European routes left vacant by the nations now engaged in war. The U.S. Merchant Marine was also at an all time low, with trained mariners difficult to find. The continued reliance on foreign ships to transport our export and import trade goods resulted in an economic crisis of a few ships charging astronomical rates.

Initially, two statutes were enacted to counter this problem: the first encouraged U.S. shipowners to fill the vacancies by authorizing "the Treasury to write war risk insurance on American-owned ships; the second liberalized the terms under which American owners might transfer vessels registered abroad to the safety of American registry." [Ref. 12:pp 38-39] With the safety of insurance underwriting, many vessels quickly transferred over, but not enough to stem the tide of rising shipping rates. Finally, the Shipping Act of 1916 came into play.

a. Shipping Act of 1916

The first of three major Acts to affect U.S. shipping, the primary significance of the Shipping Act of 1916 was the establishment of the United States Shipping Board. Consisting of five members, the board was endowed with investigatory, regulatory, as well as administrative powers, "for the purpose of encouraging, developing and creating a naval auxiliary and naval reserve and a merchant marine." [Ref. 11:p 14] The secondary significant mark made by the Shipping Act of 1916 was a wartime shipbuilding program initiated by the board. Once the United States entered the fray, the need for moving a vast amount of tonnage over to Europe became even greater. By the end of WWI, the United States had, through building, confiscation and seizure, established the world's largest postwar merchant marine, over half of which was owned by the government. [Ref. 12:p 40] However, the government was not prepared or allowed to administer its own commercial fleet. This motivated the enactment of the Merchant Marine Act of 1920.

b. Merchant Marine Act of 1920

Written for the primary purpose of empowering the Shipping Board to sell off the government's fleet, this Act began with a preamble that established U.S. shipping policy which is still in effect 71 years later:

> It is necessary for the national defense and for the proper growth of its foreign and domestic commerce that the United States shall have a merchant marine of the best equipped and most suitable types of vessels sufficient to carry the greater portion of its commerce and serve as a naval or military auxiliary in time of war or national emergency, ultimately to be owned and operated privately by citizens of the United States; and it is hereby declared to be the policy of the United States to do whatever necessary to develop and encourage the maintenance of such a merchant marine. [Ref. 13:pp 4-5]

The high shipping rates of the war years soon gave way to the low rates of the mid-1920's and beyond. The government was left with most of its fleet unsold, and was also receiving strong pressure from the private sector to dispose of the remaining ships by the quickest means available. Even the Shipping Act of 1920 seemed to be aligned against the Shipping Board as it left no other outlet for ship disposal except privatization. The government was not to be in the commercial shipping business!

Eventually, the problem was somewhat resolved by what has come to be known as the Postal Act of 1928. This Act created the requirement for all U.S. mail to be transported onboard U.S. flag ships only, thus sweetening the pot for private shipowners. The 1928 Act was meant to be a 'hidden' subsidy, but poor administration and rampant irresponsibility eventually led to the passing of the Merchant Marine Act of 1936. [Ref. 12:pp 43-47]

c. *Merchant Marine Act of 1936*

Prior to the Act of 1936, President Roosevelt addressed the Congress in a message which stated in part:

> In many instances in our history, the Congress has provided for various kinds of disguised subsidies to American shipping. . . . I propose that we end this subterfuge. . . .
>
> An American merchant marine is one of our most firmly established traditions. It was, during the first half of our national existence, a great and growing asset. Since then, it has declined in importance and value. The time has come to square this traditional ideal with effective performance. [Ref. 14:p 46]

The resultant legislation, following much debate and compromise, was nationalistic in its content and revolutionary in its outlook. Expanding on the 1920 Act, the Merchant Marine Act of 1936 stipulated that all the domestic and a "substantial portion" of the foreign commerce should be carried onboard U.S. flag ships. Additionally, all U.S. flag vessels were to be built in U.S. shipyards and manned by U.S. citizens, although these two provisions only applied to subsidy programs. The final stroke for nationalism came with the requirement that everyone was to use U.S.

manufactured goods to the maximum extent possible. [Ref. 12:pp 49-50] As the 1936 preamble states:

> It is necessary for the national defense and development of its foreign and domestic commerce that the United States shall have a merchant marine (a) sufficient to carry its domestic waterborne commerce and a substantial portion of the waterborne export and import foreign commerce of the United States and to provide shipping service on all routes essential for maintaining the flow of such domestic and foreign waterborne commerce at all times, (b) capable of serving as a naval and military auxiliary in time of war or national emergency, (c) owned and operated by citizens of the United States insofar as may be practicable, and (d) composed of the best equipped, safest, and most suitable types of vessels, constructed in the United States and manned with a trained and efficient citizen personnel. It is hereby declared to be the policy of the United States to foster the development and encourage the maintenance of such a merchant marine. [Ref. 12:p 369]

Although strongly worded, with clear Presidential backing, this work alone was not enough to provide the requisite amount of tonnage for our nation's entry into WWII. Again we found ourselves behind the power curve, struggling to catch up.

2. World War II

Though not as desperate for sealift assets as before WWI, the numbers were still low. In 1939 the U.S. flag fleet carried only 14 percent of the world's commercial tonnage.[1] At the end of the war 60 percent was being transported onboard U.S. flag ships. (Much of this increase was due to the

---

[1] "Currently, in liner trades, 82% of American commerce is carried in foreign bottoms with 18% in American owned and crewed ships." [Ref. 15: p 5]

vast losses experienced by our allies' merchant fleets during the war.) [Ref. 12:p 81]

By the end of the war, however, the Maritime Commission (a follow on to the Shipping Board), and therefore the government, had control of, and responsibility for, over 5,000 ships. Most of these had been built in one of the greatest industrial undertakings ever accomplished. "At their peak production U.S. shipyards were operating at a rate which would have reproduced the *entire prewar tonnage* of the U.S. merchant marine in only sixteen weeks and the entire world fleet in less than three years." (added italics) [Ref. 12:p 74] The very fact that we needed such an undertaking is a warning that, at least for national security purposes, the United States must have a strong merchant marine at all times. There is another warning inherent within this piece of history – only by having a large industrial base within our shipyard industry, were we capable of producing over 5,000 ships in a three-year time span. Today that industrial base has shrunk to what should be an alarming point. Without viable actions taken now to alleviate this decline, we may soon face an untenable situation: The inability of the U.S. industrial base to produce new construction or even maintain the current commercial and Navy fleets.

Following WWII, the Maritime Commission was again tasked with disposing of the government's merchant fleet. Due to the expanded foreign markets, U.S. shipping companies were able, and willing, to purchase many of the 4,000 ships for sale. Most of the remaining ships were sold to foreign governments, helping to replenish their depleted fleets. A small percentage were retained by the U.S. government and used to initiate the National Defense Reserve Fleet (NDRF). [Ref. 12:p 84]

3. Vietnam War

On October 1, 1949, the Military Sea Transport Service (MSTS), later to become the Military Sealift Command, was formed from the Naval and Army Transport Services. This was closely followed by America's entrance into the Korean conflict in June of 1950. MSTS, using its own fleet of 174 vessels and adding 400 ships from the NDRF as well as chartering privately owned vessels, began a coordinated effort to supply and resupply our deployed troops. [Ref. 11:p 20]

Again, the end of this war was followed by the disposal of most of the government-owned ships. However, our entrance into the Vietnam conflict in the early 60's quickly called for another increase in the size of the MSTS fleet. By 1969, when troop strength reached its highest point, the MSTS fleet numbered 436 ships, 176 of which had come from the NDRF. In total, 95 percent of all military cargo delivered to South Vietnam was transported by strategic sealift. [Ref. 11:p 21]

While general cargo tonnage was easily provided, two major difficulties were experienced by the logistics personnel. First, the availability of pier and storage space was limited. Until the U.S. armed forces literally built their own ports, the only ports with deep-draft piers were Saigon, which had ten deep-draft berths, and Cam Rahn Bay, which had a small two-berth pier. Unfortunately the Saigon piers were under the control of the Republic of Vietnam's port authority and many problems were encounter in trying to deal with these civilians. At the same time, because the build up in 1965-67 was so quick and uncoordinated, little or no storage/holding space could be found. With the addition of eight more deep draft ports by December 1967, "the average time a deep draft ship waited for a berth in Vietnam ports (went) from 20.4 days during the most critical period of 1965 to the 1970 average of less than two days." [Ref. 16:pp 8-25]

The second difficulty encountered was with the delivery and storage of fuel oil. Again, lack of storage was a factor, although in this case it was lack of secure storage. Tank farms were often the subject of mortar and other hostile attacks; the most significant loss taking place at the Nha Be facilities in 1965. The amount of oil lost was worth close to $3.5 million (1965 dollars). As the American build up continued, the demand for oil did the same, with three worldwide suppliers receiving contracts (Esso, Shell and Caltex), but even they could not, or would not, furnish the required amounts. Basically, "(t)he large volume, the long supply lines, and the uncertainties in schedules resulted in significant disruption in the worldwide tanker schedules of the three in-country contractors. Consequently, they were unwilling to commit more tankers than were absolutely necessary to support the military requirements." There is no record, however, of the military ever running out of fuel. Eventually, some tankers were used as floating storage tanks. The first tankers were provided by the commercial companies, and were later replaced by tankers from the MSTS fleet. [Ref. 16:pp 76-77]

Essentially, neither of these wars presented any real difficulty with regard to Strategic Sealift. The MSTS and commercial fleets retained enough assets to handle all requirements. Though vast amounts of cargo were moved, the relatively young age of the NDRF, along with a fairly healthy

Merchant Marine industry, led to successful build up and sustainment operations. The withdrawal was no less successful, though somewhat ignominious.

What followed upon the heels of the withdrawal was, to say the least, devastating to our national security interests. Quoting from the four letters written by Senator Denton in his commission's reports to the President: "In its first report the Commission concluded that there is a 'clear and growing danger to the nation's security in the deteriorating condition of America's maritime industries.'" Senator Denton continued in his fourth, and final report: "Both our strategic sealift capability and our shipyard mobilization base today fall significantly short of defense requirements, and we (the commissioners) fear that without decisive action the situation will worsen substantially by the year 2000." [Ref. 17: cover letter]

C. CONCLUSION

If all the previously cited reports, especially the Commission on Merchant Marine and Defense reports, are to be believed, then a contingency operation such as Desert Shield should have been a failure. Much of the government's fleet was either spread around the world, requiring long lead times to return to the U.S. for loading, or was sitting inactive with little maintenance, awaiting reactivation by crews who were no longer available. The Merchant Marine industry was continuing to decline with fewer and fewer ships every year, many shipyards going out of business, and the average age of the active mariners was steadily rising into the 60's. In essence, had a number of events not taken place, the sealift of equipment and supplies to Saudi Arabia would have been hopelessly overwhelmed and unaccomplished.

III. DESERT SHIELD/DESERT STORM

A. SUCCESSFUL, BUT—

Operations Desert Shield and Desert Storm have provided an excellent scenario for analyzing our current Strategic Sealift capabilities. The outstanding success of these two operations, however, should not be used as a reason for dismissing, or ignoring the deficiencies still present within our strategic sealift program.

1. Analysis

Today, even with the Fast Sealift Ships, the Ready Reserve Force (RRF) and the many chartered ships in the government's fleet, the United States does not have a sufficient level of the *necessary* (militarily useful) ships and *qualified* personnel to handle a full scale contingency operation. Neither are these ships and personnel readily available in the U.S. commercial sector. On the surface this seems contrary to the remarkable outcome of Operations Desert Shield and Desert Storm. However, a closer inspection of the facts indicates that, in these two operations and for a number of different reasons, the United States "got lucky."

2. The Facts of Desert Shield

For many years key decision makers in Congress, and the Navy for that matter, have either ignored or failed to act upon, the numerous warnings regarding the declining Merchant Marine industry. This includes warnings from the Navy's own Military Sealift Command (MSC), regarding our true ability to support the surge and sustainment of war or emergency situations. Although this historical 'apathy' is not the exclusive cause for the deficiencies, it has been a major factor.

a. Phase I

As the renamed Military Sea Transportation Service, the Military Sealift Command has been given the responsibility to provide ocean transportation for all services, as well as other U.S. Government agencies, during peacetime operations. During war or national emergency situations, MSC is the operational commander for all strategic sealift. Combatant command was assumed by the newly formed United States Transportation Command on August 1, 1988. [Ref. 5:p 7] Therefore, when the President's call went out, MSC was the one to answer for strategic sealift operations.

As of 1 August 1990 the active Strategic Sealift force consisted of: 11 chartered U.S. flag dry cargo ships, 26 chartered U.S. flag tankers, and a total of 25 chartered prepositioned ships. In the Ready Reserve Force were 83 dry cargo ships, 11 tankers and two troop ships. Additionally, on the East Coast there were eight converted former cargo ships, now known as Fast Sealift Ships, and, on each coast, a former tanker, now a hospital ship (T-AH). [Ref. 4]

According to MSC, and prior agreements with the Department of Commerce (MARAD), the order of call up for these ships in a war or emergency situation would be: [Ref. 4]

    1st - U.S. flag ships already under charter to MSC/U.S. government

    2nd - Prepositioned ships, both Afloat and Military

    3rd - Fast Sealift Ships (SL-7s)

    4th - Contracted space on U.S. flag liners

Body-29

5th - Additional U.S. charters (including foreign flag)

6th - Requisitioning: not implemented for Desert Shield/Storm

(1). Surge Sealift

On 2 August 1990 Iraq invaded Kuwait and on 7 August President Bush ordered the first troops to the Persian Gulf Area of Operations; thus 7 August became "C" day. The initial surge sealift was provided by launching two of the three Maritime Prepositioning Squadrons (MSPRONs), ten ships from the Afloat Prepositioning Fleet, seven of the Fast Sealift Ships and an activation call for ships from the Ready Reserve Force.

Due to the MSPRONs deploying from Diego Garcia (British Indian Ocean Territory), Guam (Western Pacific) and Saipan (Western Pacific/North China Sea) the first heavy combat equipment was delivered to the Gulf by C + 8 (15 Aug); a total of 105,000 short tons. The MSPRONs alone carried the entire amount of equipment and supplies to support the 1st and 7th Marine Expeditionary Brigades for 30 days. [Ref. 4] and [Ref. 18]

Also deployed from Diego Garcia, as well as from the Mediterranean and the United Kingdom, were ten of the 12 Afloat Prepositioned ships (eight cargo and two tankers) carrying "bare base, sustainment, medical supplies and port-support equipment for the Army, Air Force and Navy." By C + 14 (Aug 21), 102,000 short tons were delivered by these ships. [Ref. 4] and [Ref. 18]

Seven of the eight Fast Sealift Ships were able to get underway and deliver their massive amounts of equipment and supplies. This included equipment and supplies for the 24th Infantry Division, the 101st Air Assault Division, the 3rd Armored Cavalry Regiment, and the 18th

Airborne Support Command. Additionally, they carried follow-on support for the U.S. Marines. As an example of these ships' large capacity, USNS CAPELLA, the first to arrive at the Port of Savannah (10 Aug 90, or C + 3) [Ref. 18], alone delivered the entire 24th Infantry Division. The 24th Infantry Division was in place by 27 August or C + 20. Ultimately, five entire U.S. Divisions were be sent (though all previous plans had been for a total of three and one-third divisions), with all the assorted necessary equipment and supplies. [Ref. 4]

At the same time, 44 ships were called to be activated out of the Ready Reserve Force (RRF) and a number of ships were chartered from the U.S. flag fleet. Though the RRF ships did not necessarily meet their expected activation times (see section 2.c. below), over 70 were eventually activated successfully. [Ref. 4] However, as had been anticipated, most of the U.S. flag fleet was engaged in transportation operations around the globe, and was thus unavailable for the initial surge operation. Additionally, many of those vessels available in U.S. ports were not militarily useful. Therefore, it became necessary to charter a number of foreign flag vessels. [Ref. 5:p 5] and [Ref. 8]

(2). Sustainment Sealift

The sustainment sealift was provided by the additional RRF ships, continuous use of the Fast Sealift Ships[1], additional chartering of U.S. flag and allied vessels, and through the use of the Special Middle East Sealift

---

[1]. Following initial deliveries, these ships continued to turn around and resupply with USNS POLLUX completing at least six voyages. [Ref. 1]

Agreement (SMESA). This agreement was entered into by many nations not normally considered allies by the United States. Under this agreement, the foreign ships, already contracted to one of four U.S. flag companies, would be used as "feeders." While normal U.S. flag routes do not include the Persian Gulf; many smaller, foreign flag vessels do service the Gulf states on behalf of their U.S. flag company. For example, a specified amount of space onboard a U.S. flag vessel would be chartered, instead of chartering the entire ship, and then the ship, on its normal route, would deliver the cargo, often to a port in Spain. [Ref. 19] and [Ref. 8] From there a vessel under the SMESA would feed that cargo to Saudi Arabia, usually the port of Jiddah. [Ref. 18]

(3) Foreign charters

The problem with chartering foreign vessels with foreign crews is threefold. First, foreign governments often control the ships under their flag. For example: "At one time, a significant portion of the EUSC (Effective U.S. Controlled Fleet, American owned, but foreign flagged) was under the control of a military officer, General Noreiga of Panama." [Ref. 15:p 4] Second, foreign maritime laws can often be used against the interests of the United States, while U.S. maritime laws would not. For example: Speaking on chartering foreign flag ships for the Gulf war, the editor of *Marine Log* stated:

> In one instance, the officers and crew of a feeder ship chartered in by American President Lines refused to take the vessel into the Persian Gulf area once war was declared. Interestingly, though documents filed with MarAd indicated that the vessel was registered in Cyprus, its registration had expired. The documents filed with MarAd indicated that the ship had German officers and a Maltese crew. Had the vessel been registered in Cyprus, these civilian mariners would have been legally entitled to refuse to enter a war zone.

> Cyprus maritime law follows English common law and appears to hold that the outbreak of war during a voyage radically alters the circumstances of the seafarers' contract with the owner. Therefore seafarers are entitled to negotiate a new contract or to hold the existing contract void. U.S. maritime law is very different. A seafarer signs up for a voyage and must stay on his ship until relieved, regardless of whether a war breaks out or not. [Ref. 20:p 3]

Third, foreign crews often do not show up, or rebel, or jump ship. For example: "In another incident, the 27 Bangladeshi crew members of the freighter *Banglar Mamatar* quit lining the holds of the vessel with plywood in Oakland, Calif., prior to loading military cargo for the Gulf, and jumped ship.... Reportedly, Sea-Land also had some problems just prior to hostilities when the crew of one foreign-flag ship refused to enter the Gulf." [Ref. 20:p 3] (See Figure 3.1 for the ratio of U.S. to Foreign Flag charters. [Ref. 18])

b. *Phase II*

On 8 November 1990 (C + 96), President Bush set in motion "Phase II" of the U.S. mobilization. As this was somewhat anticipated, many of the necessary ships were already in motion, awaiting orders. Eventually, following their call-up priority list, MSC contracted for space on U.S. flag liners. A total of 19 dry cargo ships and 26 tankers were ultimately chartered. This was followed by the chartering of allied vessels. The result was a total of 73 dry cargo ships and 12 tankers chartered. Also, under the Special Middle East Sealift Agreement (SMESA), 108 foreign flag vessels provided service for the Persian Gulf operations. The last ships to be called upon were the RRF ships. By the end of the operations, 50 dry cargo ships and 21 tankers were activated out of the RRF. [Ref. 4] (See Figure 3.2 for a breakdown of ships committed to Desert Shield as of 22 February 1991. [Ref. 18])

OPERATION DESERT SHIELD/DESERT STORM
U.S. AND FOREIGN FLAG CHARTERS

A/O 4 March 1991

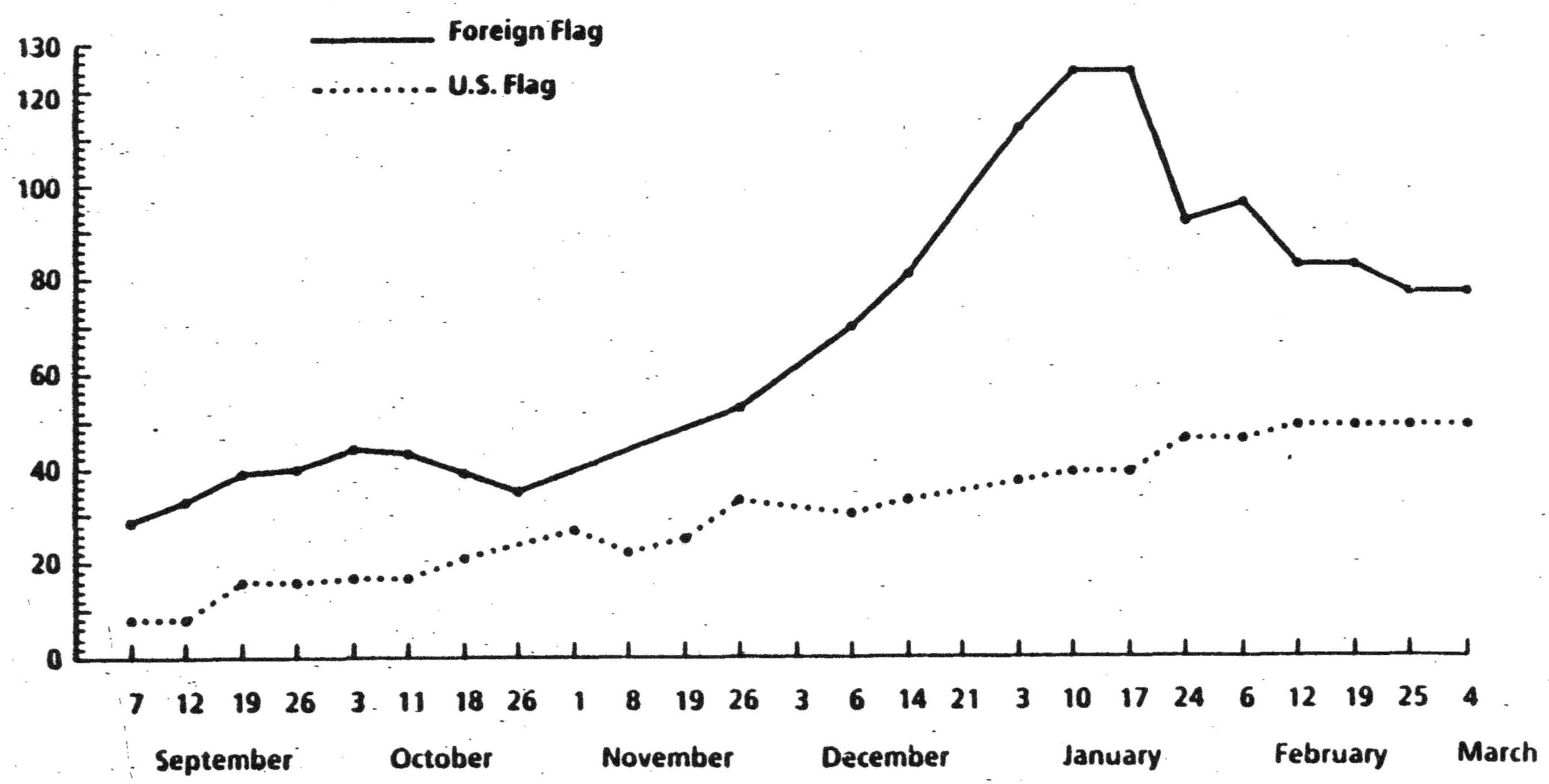

Figure 3.1

SHIPS COMMITTED TO DESERT SHIELD

As of 22 February 1991

FSS 8

RRF 76/98 Activated
73 Tendered to MSC

- RORO 17/17
- LASH 4/4
- TACS 5/8
- OPDS 2/2
- BREAKBULK 37/49
- SEABEE 3/3
- SEATRAIN 2/2
- TAVB 2/2
- TANKERS 4/9

20 Breakbulks activated in December

CHARTERS 152

- U.S. Flag 49
- Foreign (No cost) 7
- MSC Control 6
- Foreign Flag 70
- Tankers 20

PREPOSITIONED SHIPS 23

- MPS 13/13
- APS 10/12

Sustainment: 10,000 40 ft. Container Equivalent Units/Month (30 sailings)

Figure 3.2

c. The RRF

The Ready Reserve Force is a special group of ships which are supposed to be maintained such that they can be activated within five-, ten-, or 20-days.[1] The reality for Desert Shield/Storm was quite different.

Initial call-ups were: [Ref. 4]

10 Aug:	17 RO/RO and 1 Break Bulk (BB)
15 Aug:	3 LASH and 2 SEABEE
18/19 Aug:	12 BB, 2 TACS and 1 SEATRAIN
29 Aug - 9 Nov:	3 BB, 1 LASH, 1 SEABEE, 1 SEATRAIN and 1 OPDS

During Phase II:

04 Dec:	17 BB
07 Dec:	6 BB and 1 OPDS

(See Figure 3.3 for the number of RRF ships activated and tendered to MSC; source. [Ref. 18])

Many dificulties were encountered on the way. For example, in the initial activation of the first nine ships called up on the West Coast, none

---

[1] In deference to the Maritime Administration, this five, ten or 20-day requirement is simply an arbitrary number assigned to each ship. There are no clear guidelines for determining just how long it should take to reactivate a ship. [Ref. 20]

OPERATION DESERT SHIELD/DESERT STORM
RRF SHIPS ACTIVATED/TENDERED TO MSC

A/O 22 February 1991

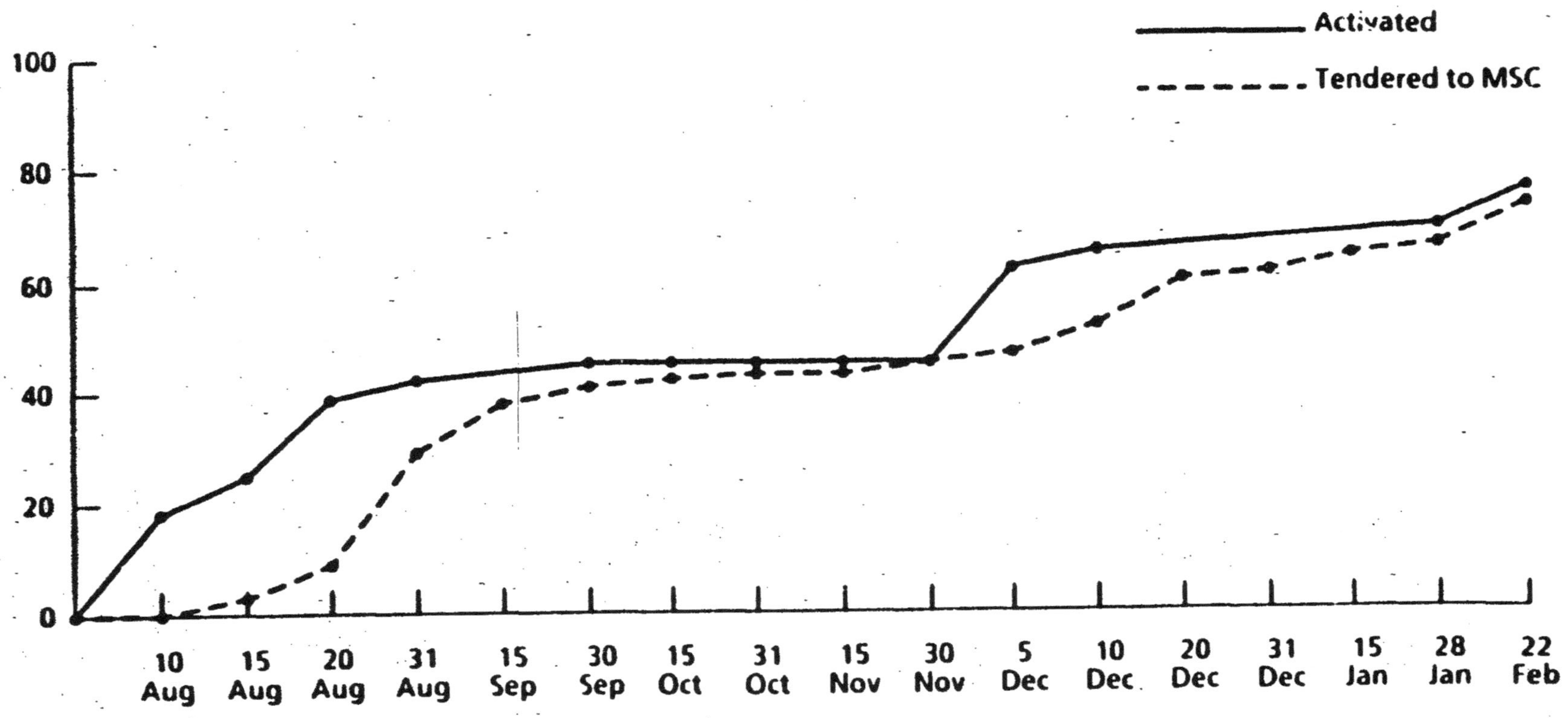

Figure 3.3

were completed within their "required" five day readiness times. (See Figure 3.4 for the overall RRF Performance. [Ref. 18]) The worst case was COMET, an ex-USNS RO/RO vessel. It took 14 days, 21 hours and 15 minutes to activate. The best case was SS CAPE BRETON, a Break Bulk type vessel. It took only five days, four hours and 52 minutes. The average time was ten days, one hour and 40 minutes, more than double the time expected. [Ref. 21:pp 16-58] The reasons for these delays varied. The major problems encountered were:

-In the engineering plants the gaskets were often deteriorated due to dehydration. Also many of the engineering personnel were unfamiliar with the plants of recently purchased foreign vessels.

-In crewing, the lack of familiarity with the ship caused many delays. The U.S. Merchant Marine was already experiencing shortfalls with its watch and engineering officers; crewing the RRF ships became that much more difficult. Some crews were slow to show up or did not show up at all.

-In the shipyards, personnel were in short supply due to the nationwide activations. The number of shipyards capable of handling the activations have dwindled considerably. Also, many of the workers were unfamiliar with the foreign-built equipment.

The most consistent problem encountered was also the worst; the condition of the vessel at breakout. If the ship had been in lay-up for an extended period of time, with no maintenance activation conducted, it was inevitable that major problems would be experienced throughout the vessel. This included berthing spaces, working spaces, galley areas and, of course, the

Operation Desert Shield/Desert Storm
RRF Performance

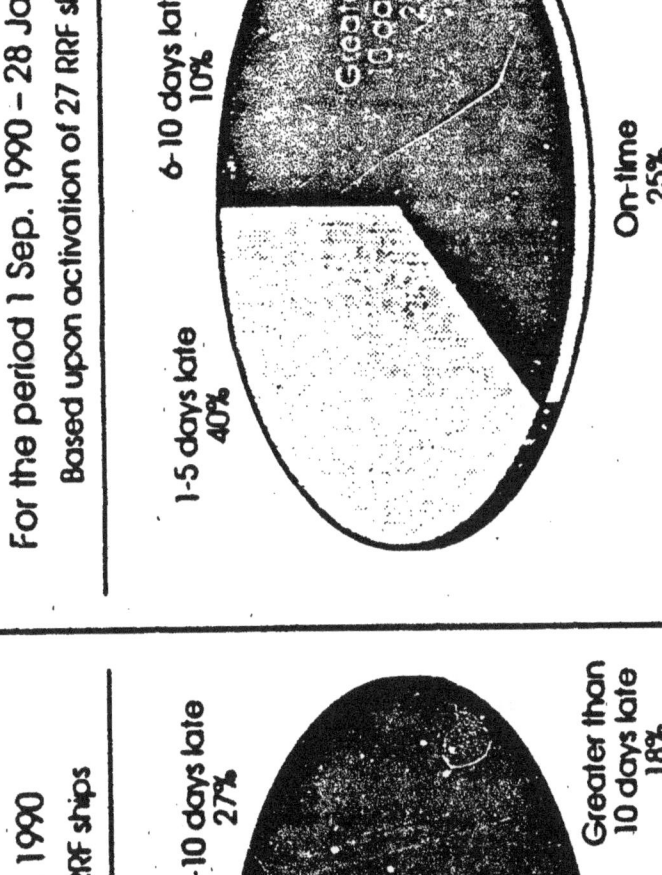

Figure 3.4

engineering plants. [Ref. 21:pp 16-58] The cutting of last year's budgeted maintenance dollars did not help the situation. [Ref. 8]

d. Additional Facts

1. Ships: The various sealift ships travelled an extremely long distance, some upwards of 8700 nautical miles. This required between 13 and 24 days of steaming time for most U.S. ships, and between 23 and 33 days for the SMESA ships.

In all: For those ships directly under MSC Operational Control; a total of 206 ships and over 600 lifts, with 3,148,884 short tons of cargo and 6,032,488 short tons of fuels delivered by 6 Mar 1991. For those ships under SMESA; a total of 108 ships and over 254 lifts, with 2700+ Forty-foot Equivalent Units (FEU's) per week delivered. These ships made at least 30 sailings per month with 643,330 short tons delivered by 6 Mar 1991. [Ref 4] (See Figures 3.5 and 3.6 for a graphical depiction. [Ref. 4])

2. Environment: The environment of the Gulf is hot and extremely dusty. This requires many modifications to normal shipboard operations, including special protection for ventilation ducts, and careful monitoring of temperatures in the engineering plants. Additionally, heat stroke and sunburn are constant dangers.

3. Ports: The ports and their facilities were outstanding. Modern deep-draft piers, with vast amounts of storage space, were available at all the different ports. This was due in large part to a modernization program conducted by the Saudis in the early 1980's.

4. Time: As already indicated, the first heavy equipment arrived within eight days of mobilization. With this auspicious beginning

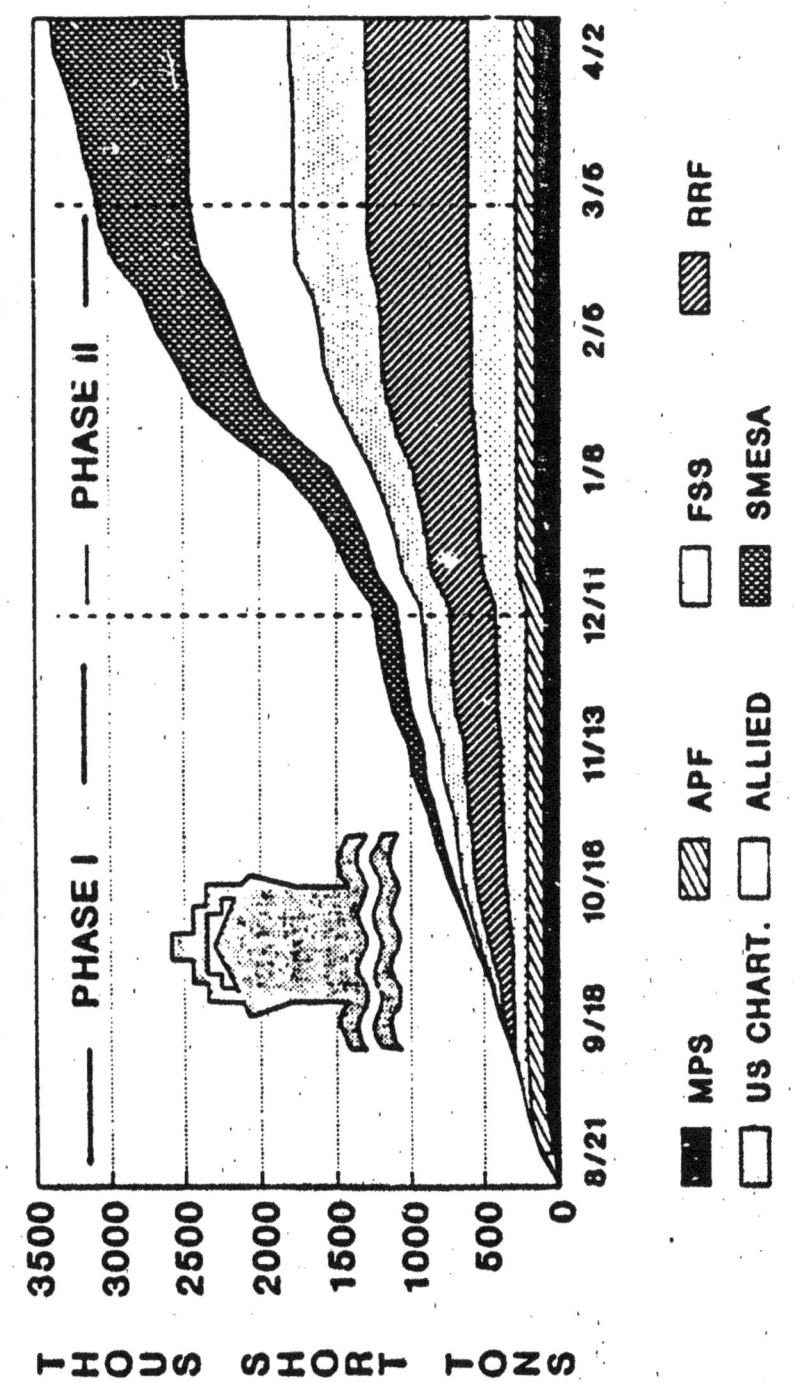

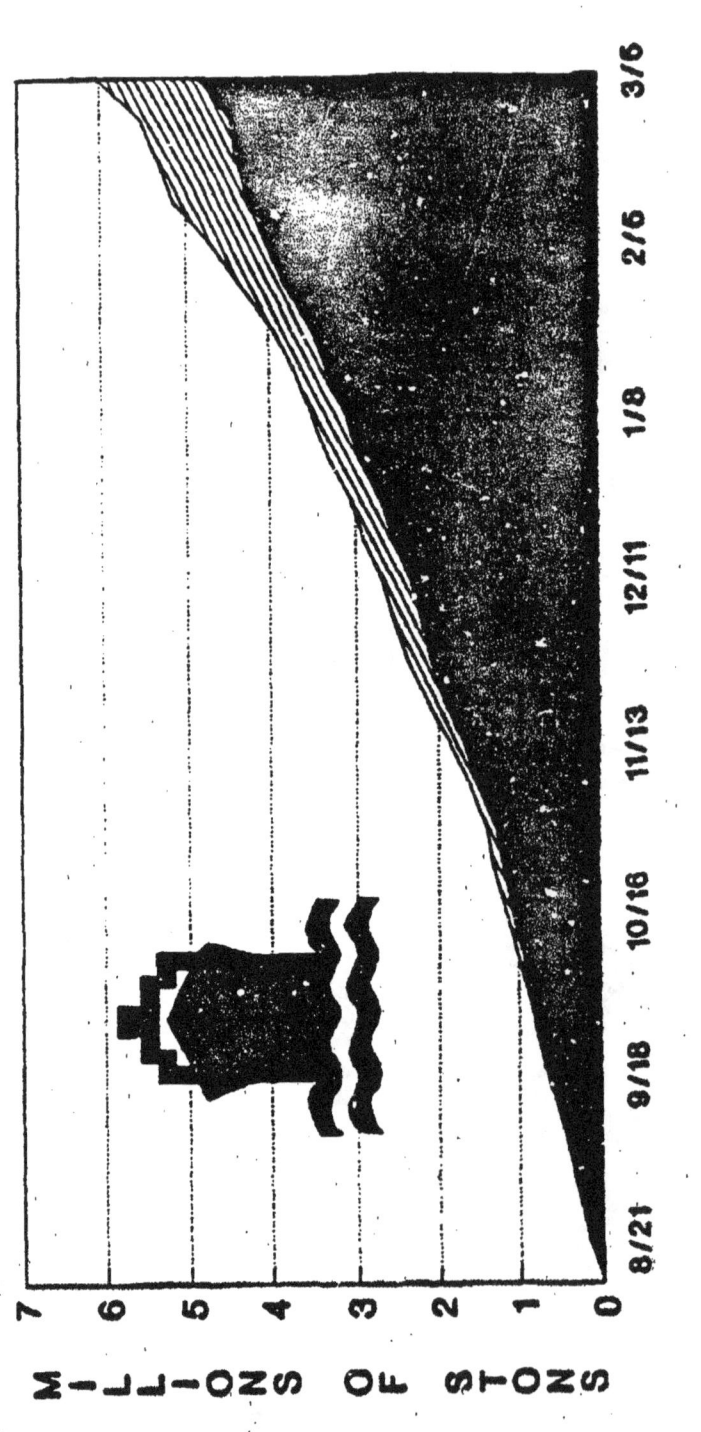

Figure 3.6

the Coalition Armed Forces were allowed, by circumstances, over five months to establish and maintain routes of supply and resupply before the first hostilities broke out. They then saw an additional month of air warfare before our ground forces advanced.

3. What Made It Successful?

a. *Time*

The United States made it, on time with some supplies to spare, but only because it "got lucky." The difficulties encountered throughout the operations were enormous. From the aforementioned delays in RRF activations, to the continuously changing movement requirements and priorities [Ref. 4], the United States was in trouble, except for one major thing - time. By not attacking the coalition forces, Saddam Hussein gave the U.S. and her allies time to prepare. Time was the critical factor.

> On 14 August the media reported, "Most of the heavy armor and other mechanized equipment needed for ground *defenses* (added italics) has only begun to be loaded on ships. Military officials said it could be 30 to 45 days before adequate ground forces are in place." On 30 August, the projection was worse, "U.S. armed forces will not have amassed a fully credible defensive force for another six to eight weeks." [Ref. 22:p 73]

With plans for 90% of the equipment and supplies to come over by ship, sealift was an expected major player. (See Figure 3.7 for sealift/airlift comparison. [Ref. 18]) Additionally, the associated long lead times to arrange, load, transport and unload the cargoes, made time a major factor. However, by the end of October the defense was in place and the Iraqis still did not attack the coalition forces. In early November, President Bush called for an additional buildup to put in place an offensive capability. Still the Iraqi's held

Body-43

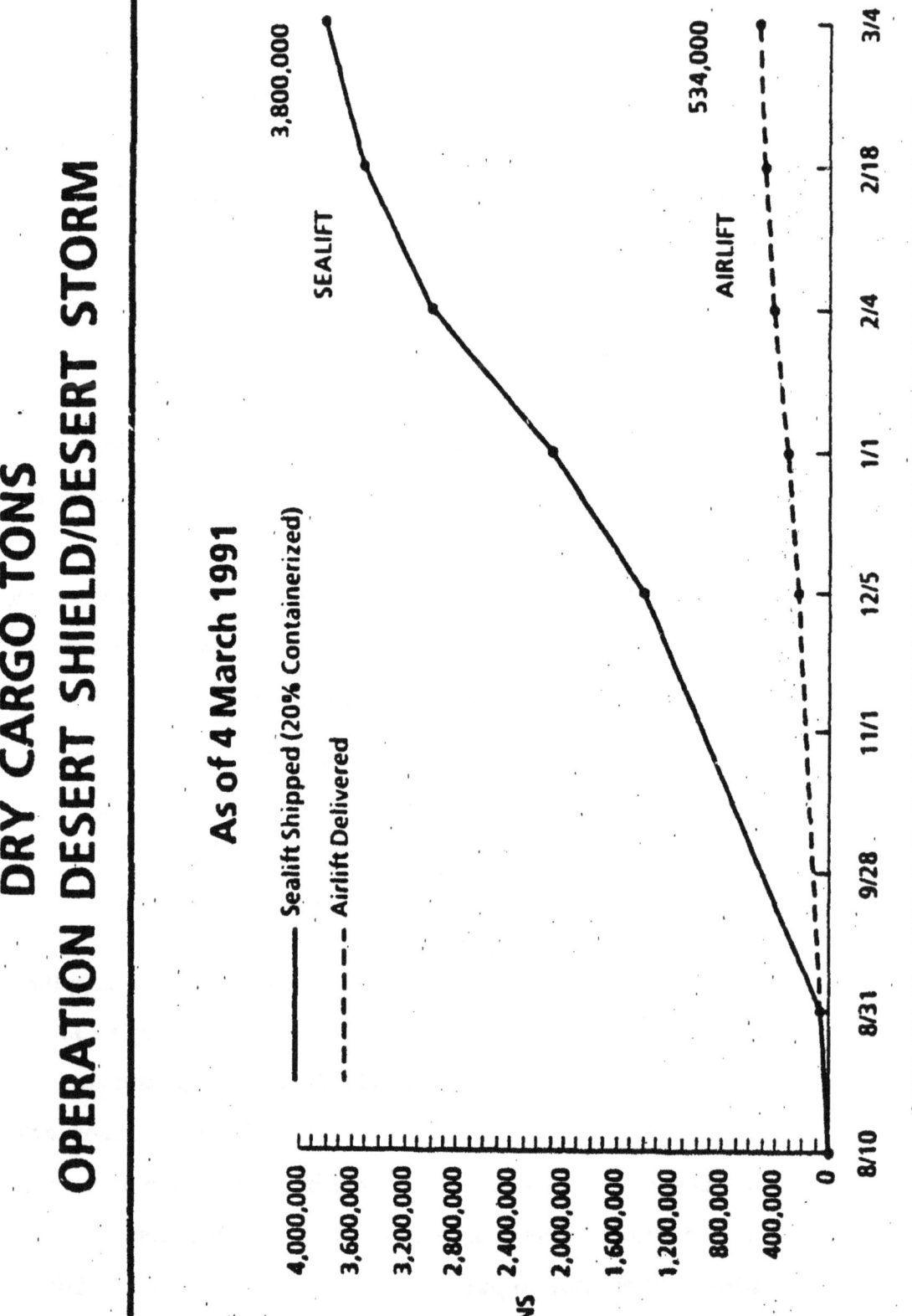

Figure 3.7

off their attack. Eventually, following the passage of many a United Nations imposed deadline, the Coalition attacked the Iraqis, first through the air and then over the ground. The time provided the coalition forces was the first 'fortuitous event,' assisting in the final victory.

b. Political Climate

Without the difficulties, trials and tribulations of the Viet Nam conflict, this author does not believe we would have seen as massive, nor as intricate a defensive, and then offensive, force put into play. Time and time again the American public was told that this would not be another Viet Nam, that everything would be done to ensure a quick and decisive conclusion to the situation. While this led to a greater demand on our sealift capabilities, it also lent itself to a stronger backing of the efforts necessary to ensure successful completion. Thus, by the infamous way it was conducted and concluded, Viet Nam played a major part in the minds of the American people, the Bush Administration and the leaders of the Armed Forces.

c. Town Criers

Without the emphasis of a few proponents (From Presidents Wilson and Roosevelt to Secretary Lehman) for a strong strategic sealift, the establishment of the RRF, the MSPRON's, the Afloat Prepositioning Fleet and the Fast Sealift Ships would not have taken place. The efforts by a few to ensure adequate strategic sealift are what made it possible to land equipment, supplies and personnel so quickly.

> In the words of Vice Adm. Francis R. Donovan, USN, Commander, MSC: "The planning and expenditures that went into strategic sealift capability over the last decade have paid off handsomely. The kind of ships we acquired for contingency sealift operations have done the job we hoped they would do." [Ref. 5:p 29]

Body-45

The only thing that might have provided an even greater success would have been a call-up for RRF activations at least 17 days prior to August 7th (C-17). [Ref. 4]

d. Special Agreement

The Special Middle East Sealift Agreement (SMESA) was, more than likely, a once-in-a-lifetime deal. The cooperation achieved was, to say the least, phenomenal; without that agreement serious shortfalls might have been felt. While the U.S. flag vessels were available, if they had had to make all the deliveries, their own regular trade routes would have suffered. This would have put a serious strain on, not only the carrier's economic status, but also on the nation's. If ships that normally pick up and deliver in the U.S. are elsewhere doing the same, then other ships must take up the slack. Under this agreement only space, not the entire vessel, was chartered, and only normal routes were taken by both the U.S. and their foreign flag ships. [Ref. 8] This was another 'fortuitous event' due in large part to the garnering of positive world opinion by the Administration and the State Department.

e. The Allies

The allied assistance was very important and necessary. As indicated above, a total of 85 vessels were chartered. The difference between SMESA ships and Allied ships is that only space onboard a SMESA vessel was chartered, while voyage and time charters were made with Allied ships. [Ref. 8] Also of note is the overall attrition rate for all fleets - 0%. None of the vessels were opposed in any manner whatsoever. [Ref. 4] *f. Environment/Port Facilities*

Without the Saudis providing a major portion of our water and petroleum requirements, all of these requirements would have had to be transported via our own strategic sealift assets. By not having to charter a vast amount of tankers, MSC was able to concentrate its efforts, and funds, on chartering cargo vessels.

The amazing and modern Saudi port facilities, built in the early 80's, were also a phenomenal blessing. They provided the ability for all the ships to make a fast turnaround, as well as have a place to store the delivered material along with the containers in which much of the material was delivered. Additionally, the required Material Handling Equipment (MHE), for off-loading equipment and supplies, was readily available in some ports. One port had just purchased brand new MHE; enough to handle the entire amount of cargo delivered there. This was an extremely important aspect. If the necessary MHE had not available then it too would have had to be transported, displacing critical military equipment on the initial sealift runs. Also, if it had not been available and had not been previously transported to the port, then debarkation would have been considerably extended or even impossible to accomplish. [Ref. 23]

g. World Shipping Market

By the time Iraq invaded Kuwait the world market for break bulk (BB) and roll on/roll off (RO/RO) ships had slumped to an all time low. With the advent of containers, these ships are no longer economically viable for commercial usage. They are, on the other hand, very useful in transporting military equipment and cargo. Many carriers, foreign and domestic, still have some of these ships in their inventory but, for ecomomic reasons, they were not being used. Thus, many were readily, and happily for the owners, available for chartering by Military Sealift Command. [Ref. 24] Even now, only one of the major U.S. flag carriers has two RO/ROs left in their fleet, with plans to soon reduce that number, while another has only one RO/RO left and will probably dispose of it soon. [Ref. 25]

h. World Situation

Finally, during President Reagan's terms in office, the U.S. Armed Forces experienced the greatest buildup since WWII. This occurred in spite of the 1984 Gramm/Rudman/Hollings Budget Act. Then, in late 1989 and throughout 1990, changing strategies within the United Soviet Socialist Republics caused the USSR to experience what can be termed 'major setbacks' in Eastern Europe. This has led to the United States setting back its own time tables (from a few days expected warning to many months) and essentially relaxing its defensive posture. Therefore, in the spring and summer of 1990, Secretary of Defense Richard Cheney placed before Congress an extensive plan for 'drawing down' our conventional forces in Europe. By November 1990 a vast majority of men, equipment and supplies were packed up, or packing up, and ready for transport home. Instead, they ended up being packed and ready to go to the Gulf. Four days after the Phase II alert order (8 Nov 90), 13 ships were berthed in Northern Europe awaiting cargo and 11 ships were berthed in the Continental United States awaiting cargo. This was due in large part to anticipation by MSC's planners; they had the ships off the various coasts awaiting orders days before the President made his announcement. For Phase

II the first ship was loaded and underway for the Gulf on 21 November 1990. [Ref. 4]

4. Conclusions

"We should not leave this exercise and say our strategic lift is OK, because it isn't. We're a victim of our success."-- Vice Admiral Paul D. Butcher. [Ref. 26:pp 10-11] Much has been done to provide the United States with the necessary strategic sealift capabilities to meet any contingency, if not war. Much still remains to be done in many areas.

In his introduction to Military Sealift Command's 1990 Annual Report, Vice Admiral Francis R. Donovan wrote:

> ... the initial stage of Operation Desert Shield clearly demonstrated the following valuable lessons:
> (1) The investment in surge sealift has proven invaluable.
> (2) U. S. flag surge sealift was inadequate to meet all Department of Defense requirements, and the charter of foreign flag ships was necessary.
> (3) There is adequate sustainment sealift through the use of U.S. flag container ships.
> (4) The adequacy of mariners to crew the ships must be carefully examined and ways considered to increase the availability of trained crews. [Ref. 5:p 6]

It is the author's opinion that the third statement is only true when looking at the senario just enacted. Only by the 'fortuitous events' previously listed was MSC able to complete its mission so successfully. For example, if the United States had been defending the state of Israel, the SMESA could not have been formed, the allies of Desert Shield/Desert Storm would have been, at the very least, neutral, and U.S. flag vessels would have been called upon to serve the entire route into the contingency area. [Ref. 8] Additionally, while sufficient tonnage does exist within the U.S. flag commercial fleet, removing those ships from their normal routes would have caused serious difficulties for both the carriers and their normal shippers.

In the recent war 27.7% of all cargo carried by American Presidents Line (APL) was in support of Desert Shield/Desert Storm. They accomplished this through the use of basically four ships, discharging their cargo in Fujaro with feeders (under SMESA) into Saudi Arabia. If a more serious situation had developed, APL would have had to shut down all other routes in order to support the sealift effort. The economics associated with a liner carrier discontinuing normal operations are devastating. The carrier loses the route to aggressive competitors, and the customers lose extablished contracts and the lower rates previously agreed upon. Also, with all assets focused on military lifts, the civilian sector must turn to foreign carriers who may or may not be willing and able to take over the routes. [Ref. 28]

In addition, cargo space is often contracted out well in advance of the actual transport. When MSC called with an urgent requirement, APL would bump previously scheduled cargo to a later date, as necessary. Currently APL is attempting to win back customers lost during Desert Shield/Storm operations, due to the necessity of breaking their agreements. [Ref. 28]

The bottom line is that the United States is still sitting with insufficient strategic sealift assets. Whatever action is to be taken to change this situation needs to be decided now, especially in light of the recent Persian Gulf events.

IV. PROPOSED SOLUTIONS

A. DIFFERENCES IN THOUGHT

This chapter is a compilation and analysis of the many differing views and ideas suggested as solutions to insufficient Strategic Sealift assets. Most of these ideas were obtained by phone and personal interviews, along with some hard copy facsimile data. None of those interviewed denies that there exists a serious problem within the Department of Defense in providing adequate sealift for war or emergency situations. The differences arise when discussing how to solve this problem.

Other than the obvious need for additional surge sealift assets, there is the need for improvement within the merchant marine industry. Ships, mariners and shipyards all require immediate attention and action to improve their current status. As the Denton Commission observed,

> Based on its study and analysis, the Commission has concluded that there is no more militarily efficient, cost effective, and reliable way to provide the majority of the military sealift requirement now and in the future than through an active United States flag merchant marine. The ships should be militarily useful and operating, engaged in peacetime in carrying commercial cargo, and manned by United States crews. [Ref. 30:p 1]

A consensus within the four major U.S. flag companies contacted was that they will survive, although they need not remain U.S. flagged. Even if the conditions within the industry remain the same, not continuing to decline, the economics of converting to foreign flag ships and crews is still very tempting. Three of the companies claim that only by incentives, or the dirty word "subsidies," will they be able to continue flagging their vessels with the Stars and Stripes. The fourth, Sea-Land, claims that "(w)e are a strong advocate to dispose of ODS (Operating Deficiency Subsidy) because we think it has contributed to the demise of the U.S.-Flag Merchant Marine." [Ref. 28]

1. Solving the Problem of Insufficient Ships

   a. *Mobility Base*

   Many people, when speaking of sealift assets, refer to a mobility base. Another way to express this is bare-bones requirements for strategic sealift. This has never been clearly defined by the DOD. Does mobility base mean number of ships or the tonnage capacity of those ships? [Ref. 29] According to Jack Helton of Sea-Land Service, Inc., "the lift capability of the surviving carriers will increase as it has for the past 10-20 years." [Ref. 28] This author agrees, and suggests that this is another reason Desert Shield/Storm was so successfully sustained. Even though the number of ships within the U.S. flag fleet has declined, the tonnage capacity has increased. Therefore, when setting down the mobility base, or bare-bones requirements, it should be expressed in terms of tonnage capacity. This makes even more sense when looking at how the civilian sector defines its requirements, by the needed space, or tonnage, and not by the ship.

   b. *Insufficient Surge Sealift*

   Another difference of opinion can be seen when examining solutions to insufficient surge sealift. The problem was stated by Admiral Donovon:

(2) U. S. flag surge sealift was inadequate to meet all Department of Defense requirements, and the charter of foreign flag ships was necessary. [Ref. 5]

With an obvious need to acquire more surge sealift assets, the Department of Defense (DOD) is looking at acquiring additional RO/RO vessels for placement into the RRF and/or the MPS. [Ref. 8] (For further discussion on this topic see section 1.c. below.) However, the civilian sector believes that they too can contribute to the surge sealift effort. As a Sea-Land representative said,

> (W)e do believe that jointly, the DOD and Industry will find a way for the container vessels and the intermodal infrastructures to contribute to the surge phase of strategic sealift to a much larger degree than experienced in Desert Shield/Storm. [Ref. 28]

A key point to note is that no "way" has yet been found to use commercial container vessels for surge sealift. Whether or not MSC and the U.S. flag carriers can find a way for the commercial ships to be used is debatable. MSC has refuted commercial involvement in surge sealift for a number of reasons. Basically, surge sealift, as stated in the Dictionary of Terms, must take place in the very initial stages of an emergency or war. It requires that equipment, heavy and outsized, and specialized material be ready at a few moments notice for transport to the affected area. The MPS and APF are clearly suited to perform this task; there is no time to wait for the return of vessels to an assigned port. As well, the current commercial fleet does not contain enough militarily useful vessels. The RRF, on the other hand, can and does provide the necessary ships within a very short time span. [Ref. 8]

c. Additional RO/ROs

As indicated above, to increase the surge sealift capacity, DOD is considering acquiring additional RO/RO vessels (Congress has already set aside $1.275 billion for this express purpose). [Ref. 31] Some in Washington feel that this could best be done by building the ships in U.S. shipyards, and then leasing them to the civilian sector. Most mariners, civilian and government, know this to be impractical for two reasons.

First, the number of vessels needed is approximately 20 more. The insufficiency, however, exists in today's government fleet, not the fleet of five to seven years from today. Even in foreign shipyards, the time necessary to build the required number of ships is considerably extended, upwards of three to five years.

Second, if these ships were in the civilian fleet, they would be spread around the globe plying their trade. Therefore, to have the 20 immediately available in U.S. ports, at least 60 would need to be acquired. This is an impossible number of acquisitions in view of today's budget constraints. Also, RO/RO's are no longer considered economically viable. Due to their large ratio of dunnage (the necessary support for cargo) to capacity, they are not suited to the commercial trades practiced today.

The better solution would be to purchase these ships, either from the U.S. flag companies ready to sell or in foreign markets. This would make the ships available now. The best solution though, would be as recommended by the Denton Commission in its Second and Fourth Reports.

> The Commission also generally supports the following ODS (Operating Differential Subsidy) reform provisions . . .:(a)

permission for a controlled number of foreign-built United States flag ships to be eligible for ODS; . . . [1]

The recommendation allowing foreign-built United States flag ships to be eligible for ODS should ultimately have no limits in terms of time, ship type, and numbers, but the Commission recognizes that, realistically, such a provision is not now desirable. The Commission therefore recommends that, for an initial year, the worldwide procurement provision be implemented, in parallel with the institution of the domestic shipbuilding programs recommended elsewhere in this report to expand the size of the militarily useful fleet . . . [Ref. 9:p 13]

That feature would allow us rapidly and economically to begin to address the acute need for more strategic sealift. Such a feature is much desired by the ship operators because it would allow them quickly to become more competitive internationally. [Ref. 17:cover letter]

This author is not necessarily supporting the ODS reform, but rather the idea of purchasing foreign-built ships and converting them to U.S. flag, all of which should be included in next year's (FY 92's) budget. Additionally, a parallel construction program in U.S. shipyards should also be added to next year's budget. This would accomplish two things. One, it would provide additional strategic sealift assets now with more in the pipeline. Two, it would provide struggling shipyards with long-term contracts that include life-cycle maintenance.

d. Additional Ships for the APF

During the recent conflict the Army encountered one of the same problems they had experienced in Viet Nam. When the accelerated

---

[1] Note: The report goes on here to recommend "(f) limited use of foreign flag feeders to improve service; . . ." as enacted for Desert Shield/Storm. [Ref. 9:p 13]

Body-55

build up began in late 1964 and early 1965, the combat troops were sent in before the logistics support personnel. This caused a lag in the logistics network, and it was not completely solved until 1967. By then the appropriate personnel, equipment and organizational commands had finally been set up and made operational. [Ref. 16:pp 13-36] This amount of time was not available to the commanders of Desert Shield. Nor will this be available for any similar situation that might arise in the future.

For Desert Shield, when the lack of logistics personnel was identified, an entire Combat Support Command (COSCOM) was ordered in early, C + 40 vice C + 90. Because of that lesson learned, the Army is now considering prepositioning all the materials and equipment for an entire COSCOM onboard APF vessels. If this idea is implemented, then additional ships would need to be procured, loaded and placed with their counterparts. [Ref. 32]

e. Operating Differential Subsidy (ODS)

Stemming from years of controversy, poor management and corrupt administrators, any subsidy is considered bad. The ODS program is no exception. It is basically a means for subsidizing the difference in cost between operating a U.S. flag vessel with U.S. unionized crews and a foreign vessel with foreign crews. In effect for many years, ODS has assisted U.S. owned companies in retaining their U.S. flag fleets.

There are two conflicting views on ODS. The first is in favor of the program. The Denton Commission, though it recommended reform, was a strong proponent of continued subsidies. The most logical reason for subsidizing the U.S. merchant marine industry is that the foreign competition is subsidized by each of their governments. In order for U.S. companies to grow and prosper in a competitive market, they need the same baseline support.

The second view of ODS is the complete opposite. As stated by a Sea-Land representative, "... we do not receive direct subsidies, ... We are a strong advocate to dispose of ODS because we think it has contributed to the demise of the U.S.-Flag Merchant Marine." [Ref. 28] In a telephone conversation with the same individual, this author requested and received an explanation for these statements. It is Sea-Land's contention that ODS has provided a "false sense of security." Consistently a leader in the industry in all areas, without subsidies, Sea-Land believes that more income can be made without relying on a subsidy. That is, if a company does not have a subsidy to fall back on then it will work much harder remain solvent. Knowing that a subsidy is available, Sea-Land believes that the other U.S. carriers have relaxed both their aggressiveness and their competitiveness. This had led those carriers to require the subsidy to survive in today's market. When questioned about Sea-Land's recent addition of capital and an intermodal network the representative explained that even though the company (CSX) which recently purchased Sea-Land was very solvent with a vast network of containers and trucks, Sea-Land itself was doing extremely well. Their, Sea-Land's, financial status was the best in the industry and they had a very good intermodal network of their own. The greatest asset provided by CSX was their financial management, not their capital. On the contrary, it was Sea-Land that provided CSX with additional capital, through the sale of Sea-

Land's assets, and with an intermodal network which far surpassed CSX's. [Ref. 29]

While the reasons both for and against subsidies seem valid, the fact remains that subsidies have been the watch-word for the Merchant Marine industry for many, many years. To suddenly remove them altogether would be devastating and, until U.S. carriers can all survive without subsidies, reforms of the existing ODS legislation are needed.

f. In Summary

Again returning to the Denton Commission,

The Commissioners pointed out some of the advantages to the nation's defense and economic welfare of obtaining strategic sealift capacity primarily from a healthy United States merchant marine:

-Active ships are immediately operational and capable of providing reliable service;

-Active ships help maintain the active industrial base needed to support an expanded strategic sealift force during time of war or national crisis;

-Active ships contribute to American economic strength;

-Active ships provide the United States with a larger voice in international rate-making conferences and ensure that foreign shipping cartels will not unfairly manipulate shipping costs;

-Active ships in a healthy United States merchant marine pay taxes and contribute favorably to the balance of payments. [Ref. 9:p 7]

2. Insufficient and Aging Mariners

a. Reserve Mariner Program

As Admiral Donovon, Commander Military Sealift Command stated,

(4) The adequacy of mariners to crew the ships must be carefully examined and ways considered to increase the availability of trained crews. [Ref. 5:p 6]

A Merchant Marine Reserve program is one of the methods being examined as a possible solution to the diminishing numbers of licensed and unlicensed marine personnel. The idea behind this program is to provide a core of mariners who would then train periodically. Then, just as the service reserves are on call during an emergency situation, so too would the merchant marine reserves be available. On the surface this looks to be a viable means to provide more seafarers. However, many questions need to be answered prior to any implementation, some of which follow.

Which agency is to administer this program? The initial proposal is for the Maritime Administration (MarAd) to manage it. This would probably be the best solution, for MarAd is already the mediator between the government and the merchant marine.

Who is going to provide the periodic training and where would it take place? Civilian companies might be willing to have additional personnel ride their ships, but compensation to the company would probably come high. MSC's fleet of ships, including the Naval Fleet Auxiliary Force, should be able to handle ship-riders, but the schedule changes, which many of these ships experience on an almost daily basis, could cause difficulties. The MPS and APF ships could be used, but their positioning so far overseas might make this an uneconomical possibility. The different Maritime academies might also provide the training, but they are pretty much limited to classroom instruction. Perhaps the best answer to this question is to marry up the program with an improved RRF readiness program. That is, assign specific reserve mariners to specific reserve ships; then on a periodic basis both mariner and ship can be exercised, including the performance of scheduled maintenance as well as full-scale vessel activations and sailings.

From where are these reserve mariners going to come? Most of the personnel in the Armed Forces reserve programs joined following active service. Special incentives, beyond those offered by the services, would need to be addressed. In other words, provide a reason to join the merchant marine reserve vice one of the service reserves. Additionally, a merchant marine reserve might end up working contrary to the basic idea of increasing the number of seafarers. It might be similar to providing these civilian mariners with an incentive to quit their six or more month a year job, find work ashore near family, receive their reserve pay and spend only a few weekends at sea.

In addition to the above difficulties, the Coast Guard is the administrator for all U.S. mariner qualification programs and has strict requirements that must be met before an individual can obtain a license. One of the main requirements is time at sea, usually measured in hundreds of days per year. A few weekends a year would not even come close to assisting in qualification or license up-grade.

b. More Ships

The more vessels available to sail, the more mariners needed to sail them. However, with the modernization of ships has come a decreasing need for crew members for each ship. Still, even one more ship in the active

U.S.-flagged fleet, with a complement of only 15, would necessitate the hiring of at least 30 more mariners.

c. Public Relations for the Merchant Marine within the Civilian Sector

Most Americans not living on or near an ocean know little about the Merchant Marine. A serious campaign geared towards improving the public's image of sailors would be beneficial to many areas. The Maritime academies are experiencing decreasing enrollments; they could only benefit from good public relations. Any mariner program should see increased interest by young men and women in going to sea. A slogan similar to "They even pay us to do this," might work well.

d. Public Relations for the Merchant Marine within the Navy

Many Navy personnel, including many officers, do not know very much about the civilian mariners. With better publicity and information, two areas of contention could be ameliorated.

First, civilian mariners are often looked down upon by Navy personnel. This is, in actuality, ridiculous because the men and women of the merchant marine are some of the finest professionals in their trade, they have to be. With a concentrated effort to improve relations between the Navy and the civilian mariners, each side could see better service and understanding from, and towards, the other.

Second, better information may increase the mariner base. By knowing what it means to be a Second Mate, a First Officer, a Third Engineer or a Chief Engineer, this author feels that many Navy officers and senior Chief Petty Officers would see the Merchant Marine as a viable continuation of their careers. Additionally, if the Coast Guard requirements for a Merchant Marine license were made known to all Navy sailors, then some may be able to complete those requirements while still on active duty. This would immediately increase the number of qualified mariners, at the same time improving relations between the Navy and the civilian mariners.

e. In Summary

Again returning to the Denton Commission,

> The Commissioners pointed out some of the advantages to the nation's defense and economic welfare of obtaining strategic sealift capacity primarily from a healthy United States merchant marine:
>
> -Active ships provide a cadre of trained merchant seamen to crew both active and/or reserve ships;
>
> -Active crews are trained and familiar with the equipment and operational requirements of the ships;
>
> -Active ships in a healthy United States merchant marine pay taxes and contribute favorably to the balance of payments. [Ref. 9:p 7]

3. Declining Shipyard Industry

The June issue of *Marine Log* provided a comprehensive look at the current status of America's Shipbuilding Industry. Sections 3.a. through 3.c. below provide quick synopses of the proposed solutions.

a. Repair versus Build

Many of the U.S. shipyards are turning to repairs vice shipbuilding. They have achieved success especially in cruise ship re or refurbishment. "But more importantly, international ship repair is far less distorted by subsidies than the newbuilding market." [Ref. 33:p 3]

b. Subsidies

Interestingly enough most commercial sources are against reinstating construction subsidies. The trend now seems to be towards taking away the foreign market shipbuilding subsidies.

> . . . Rep. Samuel Gibbons (D.-Fla.) has now introduced the Shipbuilding Trade Reform Act of 1991. It is intended to penalize all ships built with subsidy after March 21, 1991—the date on which hearings on the issue were held by the house Ways and Means Committee's Trade Subcommittee, which Rep. Gibbons chairs. . . it includes repair and conversion as well as newbuilding. [Ref. 34:p 7]

> Essentially, the bill would bar any ship ordered since March 21, 1991, and built with the aid of subsidy from loading or discharging in any U.S. port. If the bill passes and is signed into law, it will have a considerable impact on shipowners' thinking when ordering ships. Few commercial owners will want to risk being stuck with a vessel that cannot serve the world's most important single shipping market. Whether this will persuade them to buy from U.S. yards, however, remains to be seen. [Ref. 35:p 40]

In addition to this proposed bill, the Shipbuilders Council of America (SCA) is putting all efforts possible into lobbying to "bring an end to shipbuilding subsidies worldwide." This won't happen overnight and, until it does, efforts are being taken within each of the surviving shipyards to pull in as much Navy and commercial work as possible. [Ref. 35:p 36-39]

c. Revamping Current Practices

"American shipbuilders have to relearn what's involved in pleasing commercial customers." [Ref. 35:p 10] "(They) must do some real marketing work, discovering niches in the international shipping market for which they can supply ships that owners can make money from." [Ref. 33:p 3]

Essentially revamping their current business practices, they might follow the lead of Trinity Marine Group, of Gulfport, Mississippi.

> "Trinity's 'secret' is to provide the best price, quality, and delivery," say spokesman A.J. Rizzo. "The company has an aggressive sales team that hears about work and doesn't stop until they get it. Trinity is also flexible enough to take on anything, including large deep-sea vessels at its Beaumont, Texas, facility." [Ref. 35:p 39]

Especially important is for United States shipyards to change the quality of the work they perform. For many years quality in the U.S. shipbuilding industry has declined along with their reputations. Improvement in quality, improvement in marketing and improvement in management would benefit any business firm, but especially U.S. shipyards.

d. More Ships

Place the contracts for the new RO/ROs and any new APF vessels in the hands of U.S. shipyards and they could only by helped by the additional work. These contracts would of necessity be for a long period of time, anywhere from three to seven years. They could also include life-cycle maintenance, a long-term contract for continual maintenance, at the same shipyard, throughout the life of the vessel. In addition to the maintenance, additional ships will require periodic repairs, leading to more work over the subsequent life of the ship.

e. In Summary

Again returning to the Denton Commission,

> The Commissioners pointed out some of the advantages to the nation's defense and economic welfare of obtaining strategic sealift capacity primarily from a healthy United States merchant marine:

-Active ships help maintain the active industrial base needed to support an expanded strategic sealift force during time of war or national crisis;

-Active ships built and maintained in United States shipyards help maintain the industrial base;

-Active ships provide the United States with a larger voice in international rate-making conferences and ensure that foreign shipping cartels will not unfairly manipulate shipping costs;

-Active ships in a healthy United States merchant marine pay taxes and contribute favorably to the balance of payments. [Ref. 9:p 7]

B. IN CONCLUSION

Various alternatives are proposed above with some of their advantages and disadvantages identified. However, to paraphrase Admiral Robert J. Kelly, Commander-in-Chief U.S. Pacific Fleet, 'more important than finding the right answers is to find the right questions.' [Ref. 36] Prior to solving the problem of insufficient strategic sealift, the right questions must be found and asked. Only then can the right answers also be found and implemented.

V. SUMMARY, CONCLUSIONS AND RECOMMENDATIONS

A. SUMMARY

In the past decade strategic sealift has been the subject of a myriad of studies and reports. Each of these concluded that the existing assets are insufficient to adequately handle a national emergency or war situation. This is no new problem, however; it has been in existence the entire century. In the 1920s and 1930s, Congress, well aware of the problem, enacted legislation designed to solve the problem, usually by some sort of assistance to the Merchant Marine Industry. These acts served to establish marine policy still in effect today, most notably the requirement for two fleets, one Navy and one privately owned. However, these acts were essentially short-term solutions and therefore the problem has persisted.

Operation Desert Shield, and its follow-on, Operation Desert Storm, highlighted this problem in a real world situation. Fortunately for the United States and the Coalition Forces, many fortuitous events took place allowing the existing sealift assets to complete their main mission, strategic sealift. The greatest of these events was provided when the Iraqi forces did not attack the coalition. This, and other fortuitous events, allowed enough time to assemble, load, transport, off-load and reassemble five entire Army divisions, two Marine Expeditionary Brigades, and all the associated support troops, supplies and equipment required. The most important lesson learned from Operation Desert Shield, though, was that the United States does not have enough surge sealift to meet all Department of Defense requirements.

Additionally, in this author's opinion, the U.S. flag sustainment sealift is also inadequate to meet many scenarios, although it was adequate for Desert Shield/Storm.

The suggested solutions to this problem are many and vary as to their applicability. Some are biased with old fears of poor management: subsidies. Some are old ideas in new forms: buy a few foreign vessel now, and at the same time contract for newbuildings from U.S. shipyards. Some have never been tried before: Merchant Marine Reserves. All have their advantages and disadvantages. All have many questions that must be asked before any legislation or implementation can be carried out.

B. CONCLUSIONS

The first and obvious conclusion has already been made. Despite the results of Operations Desert Shield and Desert Storm, the United States remains with an insufficient number of strategic sealift assets. The second basic conclusion is that if changes to the Merchant Marine are to take place they need to happen now, before the industry reaches a point of no return. The third is that there exists no clear definition on how strategic sealift is to be quantified, whether by ship size, by number of ships or by cargo capacity. Before any other decisions on acquisitions can be made, the government needs to know how much to acquire.

1. Scenario Planning

There was no clear game plan for the use of strategic sealift in the Saudi Arabian area of operations. Also, the plan which came the closest was not flexible enough Desert Shield/Storm. This was evidenced by the difficulties encountered with continually changing cargoes, changing ports and even changing the order units to being shipped. Additionally, the load priorities which had been established were essentially ineffective due to senior unit commanders overriding the load plans of junior loading officers. This led to a great deal of confusion and many delays.

2. **Additional Sealift**

The surge sealift for Desert Shield was inadequate. Therefore, there exists an immediate need for additional surge sealift. If a national emergency can spring up with the suddenness of Desert Shield, five days from invasion to call-up, then waiting to acquire more sealift is ignoring history. Acquiring additional sealift can be done by purchasing or by building, or both. However, building RO/RO vessels and then attempting to charter them to the commercial sector should not be an option. They are not revenue producing, and there is no evidence that they ever will be profitable.

3. **Ready Reserve Force Readiness**

The readiness of the Ready Reserve Force is questionable. Whether this is due to unrealistic expected activation times or a lack of maintenance funds the result reads the same; too few were ready on time, and too many had major obstacles to overcome.

4. **Prior Investments**

If the U.S. had not made the investments into surge sealift in the 1980's, MPS/APF/Fast Sealift Ships and additional RRF vessels, the initial arrivals for Desert Shield would have been considerably delayed. This only highlights the need for additional surge sealift assets.

C. RECOMMENDATIONS

1. Commission on Merchant Marine and Defense

It is recommend that a close reexamination of the Commission's findings of fact, conclusions and recommendations be made immediately. These four reports are invaluable in their content and deserve a concerted effort toward implementing their recommendations.

2. Sealift Lobby in Washington, D.C.

It is recommend that MSC, MarAd and the U.S. flag carriers join in making sealift an important issue for this Administration. With the upcoming election year this is more important than ever. If no actions are taken within the year to improve the situation in surge sealift, the RRF and the Merchant Marine, then no action will be taken until after the general election. Leaving everything status quo has the potential for future disaster.

3. Maritime Administration

It is recommend that the Maritime Administration (MarAd) become a stronger advocate for the Merchant Marine. They are the intermediary between the government and the civilian sector. With their fingers on the pulse of the merchant marine industry, they are in the best position to ensure that improvements are programmed, planned and implemented. Most importantly, the MarAd needs to develop its own game plan on how they can assist the U.S. flag companies to remain U.S.-flagged.

4. Better Understanding

It is recommend that a program towards better understanding be initiated within the Military Sealift Command, the surface Navy, the government and the U.S. flag carriers. Each of these players has a number of biases and prejudices inherent within their organization. If any lasting improvements are going to be made, they must be planned and implemented with the cooperation of all players. Only through better understanding of each other will the necessary cooperation emerge.

5. **Sealift/Transportation Pipeline**

It is recommend that the U.S. Navy examine the possibility of a program specifically designed to educate a cadre of personnel in the area of sealift transportation, for eventual use in a sealift or transportation designator. While the Transportation Management curriculum at the Naval Postgraduate School in Monterey, California, is excellent, it is specifically designed to fill certain P-coded billets, rather than be a pipeline to a transportation designator. This author understands the difficulties associated with creating another Naval designator, however, if Strategic Sealift is to remain the fourth mission of the Navy, then the Navy needs to fully support it with a highly trained core of professional personnel.

6. **Strategic Sealift Plan**

It is recommend that the coordinated plan of action, following the lessons learned in Desert Shield and Desert Storm, include the identifying of actual Department of Defense requirements and placing the appropriate priorities thereon. Additionally, this plan should include an intensive examination of the very real possibility that good ports and easy off-loads will not be available in all situations.

7. **Army COSCOM onboard APF**

It is recommend that the Army follow through with a plan to place a complete COSCOM onboard newly acquired APF vessels. While this would place additional burden upon surge sealift, it would allow the necessary support personnel to be available almost immediately. It would also free up the currently required sustainment assets.

8. Future Study

It is recommend that future study be made into the following areas:

- The use, by the Department of Defense, of an intermodal network combined with Just In Time practices. U.S. flag carriers are seeing intermodalism as the watch-word of the future. A study should be made into the possibility of adapting an intermodal structure to strategic sealift requirements. Along the same lines, Just In Time provides today's businesses with up to the minute information and easy off-load access. This too should be studied for adaptability in parallel to the former recommended study.

- The possibility of establishing agreements, similar to SMESA, before any hostilities break out. That is, establish agreements now in key parts of the world which are volatile with a potential for hostilities, and look into the possibilities for other areas as well.

- The establishment of a Merchant Marine Reserve needs to be examined in depth. There are many obstacles associated with this kind of program, a few of which were identified in this thesis.

LIST OF REFERENCES

1. Office of the CNO, Strategic Sealift Division, *Strategic Sealift Program Information*, OP-42, 16 April 1985.

2. McFarland, J.M., *An Examination of the Outporting Ship Program Implemented in Response to the Increased Program Size of the Ready Reserve Force*, Master's Thesis, Naval Postgraduate School, Monterey, CA, June 1988.

3. Tippett, Louis M., Military Sealift Command, ltr to author, Subj: Response to Request for Information, Citing Joint Pub 1-02, 6 May 1991.

4. Fields, R, COL, USA, "Desert Shield/Desert Storm Briefing, as of 6 Mar 91," Washington, D.C., March 1991.

5. Military Sealift Command, Department of the Navy, *1990 Annual Report*, Government Printing Office, Washington, DC, 1991.

6. Center For Naval Analyses, Report CRM 88-234, *An Assessment of Activation Testing for The Ready Reserve Force*, by D. C. Mach and E. S. Cavin, February 1989.

7. Military Sealift Command, Louis M. Tippett, Letter to the author, 6 May 1991.

8. Interview between Col. Rick Fields, Military Sealift Command, and the author, 13 June 1991.

9. Commission on Merchant Marine and Defense, for the President of the United States, *Second Report; Recommendations*, December 31, 1987.

10. Woodrow Wilson, "Message to the Congress," cited by Samuel A. Lawrence, *United States Merchant Shipping Policies and Politics*, The Brookings Institution, 1966.

11. Whitehurst, Clinton H., Jr., *The Defense Transportation System*, American Enterprise Institue for Public Policy Research, 1975.

12. Lawrence, Samuel A., *United States Merchant Shipping Policies and Politics*, The Brookings Institution, 1966.

13. Esposito, Daniel N., *The National Defense Reserve Fleet and Strategic Sealift: Past, Present, and Future*, Master's Thesis, Naval Postgraduate School, Monterey, California, December, 1989.

14. Franklin D. Roosevelt, "Message from the President of the United States," House Document, cited by Samuel A Lawrence, *United States Merchant Shipping Policies and Politics*, The Brookings Institution, 1966.

15. Seiberlich, Carl J., RADM (Ret.), "American Merchant Marine: The Supply Line to National Security," Paper provided by American President Companies, Ltd., 30 May 1991.

16. Heiser, Joseph M., Jr., LGEN, USA, *Vietnam Studies: Logistic Support*, Department of the Army, Washington, D.C., 1974.

17. Commission on Merchant Marine and Defense, for the President of the United States, *Fourth Report; Recommendations*, January 20, 1989.

18. Garner, David, CAPT, USN, "Operation Desert Shield/Desert Storm," briefing presented to Mitre Corporation by Logistics Management Institute, 15 May 1991.

19. Interview between Bob Elwell, Mobilization Plans, Studies & Analysis, Military Sealift Command, Washington, DC, and the author, 06 June 1991.

20. Blenkey, Nicholas, "These colors sometimes run," *Marine Log*, March 1991.

21. Kessler, Phillip R., *RRF: West Coast Activations in Support of Operation Desert Shield*, Master's Thesis, Naval Postgraduate School, Monterey, CA, March 1991.

22. "Nobody asked me, but . . .," by LtCol Ky L. Thompson, *Proceedings*, January 1991.

23. Monroe, David, CAPT, SC, USN, NAVMTO briefing to Administrative Sciences students at the Naval Postgraduate School, 11 April 1991.

24. Telephone conversation between Mr. Louis M. Tippett, Military Sealift Command, and the author, 19 March 1991.

25. Telephone conversation between Mr. Jay Brickman, Crowley Maritime, and the author, 3 June 1991.

26. Hall, Kevin G., "Bush orders DOT to prepare transport in case war erupts in Persian Gulf," *Traffic World*, January 14, 1991.

27. Telephone conversation between RADM Carl J. Seiberlich, USN (Ret.), American President Companies, Ltd., and the author 11 June 1991.

28. Sea-Land Service, Inc., Mr. Jack D. Helton, Letter to the author, 14 June 1991.

29. Telephone conversation between Mr. Jack D. Helton, Sea-Land Service, Inc., and the author 18 June 1991.

30. Commission on Merchant Marine and Defense, for the President of the United States, *First Report; Findings of Fact and Conclusions*, September 30, 1987.

31. Telephone conversation between Mr. Louis M. Tippett, Military Sealift Command, and the author, 3 April 1991.

32. Fields. R, COL, USA, "Desert Shield/Desert Storm Briefing, as of 13 Jun 91," given at the Naval Postgraduate School, Monterey, CA, 13 June 1991.

33. Blenkey, Nicholas, "Can old shipbuilders learn new tricks?," *Marine Log*, June 1991.

34. Washington Update, "Gibbons Bill Covers Repairs, Too," *Marine Log*, June 1991.

35. Mottley, Robert, "American shipyards face the crunch," *Marine Log*, June 1991.

36. Kelly, Robert J., USN, CINCPACFLT, Commencement address to Naval Postgraduate School June 1991 graduating class, June 20, 1991.

INITIAL DISTRIBUTION LIST

		No. Copies
1.	Defense Technical Information Center Cameron Station Alexandria, Virginia 22314-6145	2
2.	Library, Code 52 Naval Postgraduate School Monterey, California 93943-5100	2
3.	Professor Dan C. Boger, Code ASBo Department of Administrative Sciences Naval Postgraduat School Monterey, California 93943	1
4.	Mr. Louis M. Tippett Military Sealift Command Department of the Navy Washington, D.C. 20398-5100	1
5.	Professor Alan W. McMasters, Code ASMg Department of Administrative Sciences Naval Postgraduat School Monterey, California 93943	1
6.	Defense Logistic's Studies Information Exchange U.S. Army Logistics Management Center Ft. Lee, Virginia 23801-6043	1

www.ingramcontent.com/pod-product-compliance
Lightning Source LLC
Chambersburg PA
CBHW081841170426
43199CB00017B/2812